THE SLEUTH
AND THE GODDESS

THE SLEUTH
AND THE GODDESS

HESTIA, ARTEMIS, ATHENA, AND APHRODITE IN WOMEN'S DETECTIVE FICTION

SUSAN ROWLAND

Routledge
Taylor & Francis Group

LONDON AND NEW YORK

First published 2015 by Spring Journal Books

Published 2020 by Routledge
2 Park Square, Milton Park, Abingdon, Oxon OX14 4RN
52 Vanderbilt Avenue, New York, NY 10017

Routledge is an imprint of the Taylor & Francis Group, an informa business

British Library Cataloguing-in-Publication Data
A catalogue record for this book is available from the British Library

Library of Congress Cataloging-in-Publication Data
A catalog record has been requested for this book

ISBN: 978-0-367-46106-5 (hbk)
ISBN: 978-0-367-46107-2 (pbk)
ISBN: 978-1-003-02694-5 (ebk)

Cover design, typography, and layout:
Northern Graphic Design & Publishing
info@ncarto.com

TABLE OF CONTENTS

DEDICATION

for Christine Downing and Ginette Paris—
inspiring women

ACKNOWLEDGEMENTS

The Sleuth and the Goddess could not have been written without the crucial support of an outstanding editor and publisher, Nancy Cater. The whole world of Jung, archetypal criticism, and depth psychology is immensely in her debt for her astonishing hard work and creative achievements over many years. I owe her and her team at Spring Journal Books an immense debt of gratitude.

I also want to acknowledge the help, inspiration, and comradeship of friends, colleagues, and students of Pacifica Graduate Institute throughout the germination of this book. In particular, I thank Keiron Le Grice, Jennifer Selig, Glen Slater, Safron Rossi, Nancy Galindo, Sue Gary, Pat Katsky, and Elizabeth Nelson. So many students, but especially those in the MA program I chair, Engaged Humanities and the Creative Life, and the doctoral program on which I also teach, Jungian and Archetypal Studies, offered helpful suggestions, feedback, and insights.

Special mention goes to the dedicatees of *The Sleuth and the Goddess*. Christine Downing and Ginette Paris have done so much to take Archetypal Psychology into the neglected area of women's imagination and embodiment. Although James Hillman's pioneering work is also foundational to this book, in subject matter and in research epistemology, this one is for the women.

Moreover key friends have helped this book into being through providing energy, chocolate, glasses of wine, and brilliant conversation. These include Wendy Pank, Christine Saunders, Claire Dyson, Leslie Gardner, Luke Hockley, Lori Pye, Jacqueline Feather, Evangeline Rand, Adrian and Kevin Campbell, Cynthia Hale, Caroline Barker, Margaret Erskine, Alberto Lima, Evan Davis, and Ailsa Camm. To these and my family, Cathy Rowland, John and Rebecca Rowland, Emma and Thomas, I owe a very great deal.

Of course I wish to mention the International Association for Jungian Studies for providing the essential ongoing research field for this work. In addition this book was informed by friends at the University of Greenwich, UK, where I used to teach.

To my beloved husband, the digital literary artist, Joel Weishaus, I owe more than I can say.

CHAPTER ONE

INTRODUCTION
Mythical Knowing And Detective Fiction

Opening: The Mystery of the Detective

Detective or mystery fiction is a haunting and haunted literary genre, in which there is always *another*, a ghostly, perhaps even divine, presence.[1] Doubleness characterizes its nature. On the one hand, it has alternate names of "detective" stories, indicating a focus on its lead figure, and "mysteries," hinting at metaphysical dimensions of plot. On the other hand, this form also has dual origins as both a peculiarly modern, and yet distinctively ancient, type of literature. Moreover, its doubling does not end here. It has evolved two main subgenres: "hardboiled" and "cozy," along with an ambivalence between seriousness (the crime is always murder) and frivolity (in its liberal invitation to humor). In addition, mystery fiction identifies itself with popular vacation reading, while its addiction to self-referentiality invites its readers to make use of their own knowledge.[2]

This book, *The Sleuth and the Goddess*, will explore detective fiction, or mystery's lingering duality, not least in adopting both terms to maintain awareness of this uncanny doubling. However, in order to offer depth, I will concentrate on work by women, in particular those of the United States and United Kingdom in recent decades. In its focus on women authors, *The Sleuth and the Goddess* is therefore not limited to women detectives. So, for example, I will include intrepid Marcus Didius Falco making an uneasy living in Ancient Rome at the behest of his author, Lindsey Davis.[3] In addition to exploring works by many writers, each chapter, including this one, will have four case studies of specific novels germane to the material.

On the other hand, apart from the inevitable appearance of Mr. Sherlock Holmes in this opening chapter, the work of male writers will be avoided.[4] I make this difficult decision for reasons of giving space to those who have been relatively unappreciated, and in order to further a particular exploration of gender that I will outline in Chapter 2. For similar reasons *The Sleuth and the Goddess* is limited to print novels and will not embark upon treacherous excursions into film, TV, or digital versions of the genre. These powerful treatments are, by definition, put together differently, usually collectively, and therefore repay study that attempts to adjudicate their variations from the printed or audio book.

Most significant for *The Sleuth and the Goddess* will be a new argument about a haunting of our (post) Christian modern era with ancient mythical structures identified as pagan goddesses. Those especially scrutinized are Athena, Artemis, Aphrodite, and Hestia. This introductory chapter will make a case for mysteries as goddess haunted in the sense of C.G. Jung's assertion that "the gods have become diseases."[5] What were named gods and goddesses in ancient times are now to be found in our modern dis-ease, and, this book suggests, lurking in spaces where the pressure of the dominant paradigm is less overt, such as in margins of popular culture. As I will explore later in this chapter, in such overlooked and undervalued domains, detective fiction embodies the repetitive strategy of myths as structuring processes of the human psyche. Yet here at the beginning, no exploration of myth in women's mysteries should fail to pay tribute to Christine A. Jackson's groundbreaking *Myth and Ritual in Women's Detective Fiction* (2002).[6] This remarkable work pioneers research into myth and mysteries, in particular making a case for them to be treated as more than expositions of a formula.

> Whodunits signify serious business in our death-repressed culture, and they deserve close scrutiny. Mysteries are not less than novels; they are full-fledged novels that happen to be mysteries.[7]

Jackson's book differs from this one in that, while her focus is on mythological themes such as rebirth, the underworld, the wasteland, and the environs of hell, this book looks toward the realm of goddesses. Part of this differing emphasis stems from Jackson's interest in Joseph Campbell's more narrative approach to myth, which bears out her thesis about the ritual quality of these works. Oriented to consider goddesses as modes of knowing and being, *The Sleuth and the Goddess* will instead investigate how women's mysteries incarnate multiple divinities in order to critique conventions. In doing so I suggest that mysteries offer something ancient to be renewed and reborn in the modern psyche.

Heroes Modern and Mythical

While the fictional detective appears to be a modern figure, "he" is also thousands of years old as the protagonist of a hero myth. Yet crucial to this book is the refusal to collapse the detective figure into the male protagonist of the hero monomyth famously examined by Joseph Campbell in *The Hero with a Thousand Faces* (1949).[8] Campbell's monomyth hero is a young man who receives a call to adventure, battles a monster successfully, and finally receives his reward from a grateful community in the form of a valuable bride who secures his new role as a leader.[9] Arguably such a myth resembles a male initiation rite at puberty more than the all-purpose development of an ego structure it is often taken to be.

Although some examples of detective fiction do indeed provide monsters in the form of grotesque criminality, the myth of the form is both more flexible and more precise. In a previous work, *The Ecocritical Psyche*, I suggested that the founding myth of detective fiction is that of the trickster, in which, most importantly, the tricky protagonist could stand for hero or villain or both.[10] Indeed, *The Ecocritical Psyche* makes use of Lewis Hyde's formidable work, *Trickster Makes This World*, to argue that, in Hyde's terms, if the trickster myth emerged from archaic times when humans learned to hunt and be hunted, then modern detective fiction preserves and exposes us to that elemental structure of consciousness.[11]

In possessing the trickster's plural capacities, including that of varying gender traits and deceit, the detective hero reveals flexible trickster traits. In planting such protean qualities throughout mystery fiction as an identifiable genre with specific characteristics, the trickster as genre is more precise. Indeed, by positing the trickster myth as foundation in human evolution's prehistory, this book suggests that it serves as bedrock for the various gods and goddesses of knowing that come after. Such a possibility also draws in detective fiction's close relative of crime fiction, often considered to begin in Western literature with Aesop's Fables, in which a trickster is profoundly at work.[12]

Given that detective or mystery fictions are always also crime stories, it matters that the reverse is not inevitably the case. Not all crime fiction involves a quest for truth that I take to be the necessary feature of mysteries. Moreover, strictly speaking the category of crime indicates a violation of law that could encompass such historically determined crimes as adultery and slander, whereas the modern mystery concentrates on the illegal killing of one or more human beings.

Most striking in the distinguishing of detective fiction from the larger category of crime fiction is the modern character of the figure questing for justice in the context of the relatively recent invention of a professional police force that protects society by upholding its laws. While Dorothy L. Sayers's Lord Peter Wimsey rightly asserts in the 1930 mystery, *Strong Poison*, that the modern sleuth inherits the medieval knight's quest for justice, the distinctively contemporary quality of the detective on a hero's quest is her or his relationship to the police.[13]

Once industrialization produced urban populations in the early nineteenth century, keeping order and solving crime could no longer be left to the military or privatized ventures.[14] The arrival of police forces marked a shift, in that society was prepared for a hero who defends the law rather than celebrating tricky lawbreakers such as the elusive Robin Hood. Moreover, a further mark of modernity in the detective hero is her or his emergence contemporaneously with, and resemblance to, psychoanalysis, as invented by Sigmund Freud (1860-1939), and later figures such as C.G. Jung (1875-1961).[15] This book will use the term "depth psychology" to stand for theories that stress the importance of the unconscious to psychic functioning.

What links depth psychology to detective fiction is a similar sense of knowledge as a *problem* with tricky overtones, rather than something subject to wholly rational analysis. Both the detective and the psychoanalyst have to search for clues to a truth that is hidden from view. In both cases the truth sought for is of unknown extent. Both cultural forms additionally structure their quests in terms of seeking an ultimately narrative understanding of knowledge. This is to say that the revelation of "whodunit" is not enough; it is the *process*, the story of discovery, that embodies the true knowing of the genre, as I shall show.

Finally, both depth psychology and detective fiction partake in what has come to be known as "modernism," a movement in the arts and sciences that by the end of the nineteenth century was challenging the dominance of rational modes of knowledge.[16] What for historians had become a complex weave of colonial nations and for cultural critics the subsequent awareness of arts quite outside Western tradition, for scientists was discoveries undermining centuries of assumptions about the structure of matter. Modernism was an attempt to reinvent or *re-figure* Western assumptions of unquestioned superiority in its rational paradigms.

In this sense, the detective and the psychoanalyst are trying to save their world from within the complex trickiness of modern culture and psyche. By aiming to find out the truth about a murder or to uncover psychic trauma, mysteries and depth psychology both emphasize the importance of what is hidden and repressed to the modern self. Consequently, they both operate mythically in seeking out those stories that embody what has been lost to the modern person.

Therefore, *The Sleuth and the Goddess* brings together four core elements: the tricky detective or mystery genre, depth psychology excavating myths, urban cultures of modernism, and goddesses as modes of knowing. It is time to look a little further at that quintessentially modern male hero of detection who later haunts writers of either gender: Sherlock Holmes.

Sherlock Holmes as Trickster and Dionysian Parent

In his very first story, "A Study in Scarlet" (1887), Sherlock Holmes meets Dr. Watson while preoccupied by discovering a test for blood. His quest for knowledge requires that he use his own body.[17]

> Broad, low tables were scattered about, which bristled with retorts, test-tubes, and little Bunsen lamps, with their blue flickering flames. There was only one student in the room, who was bending over a distant table absorbed in his work. At the sound of our steps he glanced round and sprang to his feet with a cry of pleasure. "I've found it! I've found it," he shouted to my companion, running towards us with a test-tube in his hand. "I have found a re-agent which is precipitated by hoemoglobin, and by nothing else."...

"Dr. Watson, Mr. Sherlock Holmes," said Stamford, introducing us...

"Let us have some fresh blood," he said, digging a long bodkin into his finger and drawing off a drop of blood in a chemical pipette.[18]

From this seminal moment two images haunt the later detective genre, both of which are worth examining skeptically: the *forensic* nature of detecting and the masculine rationality of the sleuth. While it is undeniable that forensic science as an ingredient of police work is seeded in this scene, in such fictions as *The Hound of the Baskervilles* (1902) Holmes also champions the *productive* method of detecting, where his renowned presence starts a reaction in a group of characters that ultimately leads to finding the guilty party.[19]

Moreover, Holmes's fixation on subjecting crimes to rational analysis is far from guaranteeing detecting as wholly masculine or predominantly rational, even for him. Living in a nurturing partnership with a man, Dr. Watson, Holmes's detecting seems to require non-rational sources such as mood-altering substances like tobacco and cocaine, and pursuit by means of myth in the eponymous "hound." Of course the resemblance in the latter story between heroic detective and deviant criminal, Stapleton, also problematizes the supposedly rational opposition of good and evil. The trickster nature of Sherlock Holmes will be examined in the following case study, but first his dissecting habits merit further exploration.

In "Dionysus in Jung's Writings," psychologist and cultural critic James Hillman points out that C.G. Jung stresses "dismemberment" as his key feature in the many myths of the god Dionysus.[20] In fact, in the distinction between Apollo, the god of light, reason, and distance, and Dionysus, divine bringer of body and ecstasy, Hillman identifies a division between the preferences of rational modernity versus the succeeding age of deviance in modernism. In Jung's dismemberment of the god, Hillman discerns a possibility of psychic renewal in the corporeal rending of an aging god. Apollonian Christianity, an era dominated by one god defined by distance and disembodiment, is to be followed by dismemberment in a mythic narrative.[21]

Jung sees a two-stage dismembering process: first comes a separation into opposites, such as the very notion of Apollonian and Dionysian itself. Opposition is then transformed into multiplicity, with a wider dispersal of the divine in matter, which both Jung and Hillman call archetypal. To Jung, archetypes are inherited potentials in the human psyche for certain sorts of images and meaning.[22] They represent the possibility of many different modes of psychic functioning; or as Hillman later puts it, a polytheistic psyche in which the gods are diverse ways of being in the world.[23] Hillman also points out that this second stage of Dionysian dismemberment entails an entry into a different type of consciousness. We enter a new cosmos with the dispersed fragments of the body of the god.[24] Distance from the divine becomes interiority within the realm of the god.

> The movement between the first and second view of dismemberment compares with crossing a psychic border between seeing the god from outside or from within his cosmos.[25]

Sherlock Holmes stabbing his own finger for blood is not necessarily going so far as to cut it into pieces. However we have previously been told of his unfeeling treatment of dead bodies in beating them to test out the possibility of post-mortem bruising.[26] Holmes does represent an attempt to stay outside the sometimes messy emotions of embodied detecting. Yet when adopting trickster methods of sleuthing, Holmes is forced to dwell within the cosmos of Dionysus, a notably tricky and ecstatic god. He becomes implicated in body and emotion. Indeed, the emerging of Dionysus may mark the evolution from the hunter-trickster myth to the Greek pantheon. Might detective fiction represent a return of how polytheism in the psyche manifests in the world? Might mysteries actually be *mysteries*?

Posited first as the binary alternative to Apollo, here Dionysus marks both a decisive break from, and a trickily, effective continuation of, monotheistic Christian culture. For if Apollo is succeeded by Dionysus as a single divine principle, monotheism is modified, not abolished. *The Sleuth and the Goddess* will argue that it is in later mysteries, by women, that the most revolutionary innovations of psyche and knowledge can be discerned. Once the Apollonian god is twice dismembered, Dionysus can recover some of "his" occluded deep nature as multiple, or polytheistic, and as feminine; in fact, as Nature herself. First of all, Sherlock Holmes's quest for Apollonian truth has to become stuck in a bog, the matter of mater, Mother Earth.

Case History (1): *The Hound of the Baskervilles* (1902) by Arthur Conan Doyle

This particular Sherlock Holmes adventure is structured like an underworld initiation with the haunted moor as the dark place at the heart of a Gothic tale. Gothic is here used in its literary sense as the term for the sensational works of the imagination that began in the eighteenth century with the ghost story, *The Castle of Otranto* (1760) by Horace Walpole.[27] Gothic continued and mutated into the modernist non-realistic genres of science fiction, fantasy, detective stories, the uncanny, horror fiction, cyberpunk, and arguably reached a dark claw into non-literary arenas such as psychoanalysis.[28]

Mysteries partake of the Gothic in their ostensible ability (more apparent than real) to tame the wilderness of the psyche and solve what challenges modern rationality. Gothic is about transgressing boundaries that are erected by the drive to rationality and its determination to remove the imagination from knowing. So divisions between reality and fiction, sanity and madness, life and death, ghost and dream are troubled by Gothic tales that struggle in their efforts to re-build the structures of reason.

The fictional detective, and Sherlock Holmes in particular, is designed to both invoke and remove the irrational by solving Gothic phenomena of the unexplained. In this sense, Holmes's dedication to science as providing a logical explanation for mysterious crimes makes him a cultural warrior, a hero of modernity. On the other hand, in *The Hound of the Baskervilles*, all of Holmes's insouciance cannot stop the spectral hound from claiming at least one victim.

The story begins and ends in London. It is on London streets that Holmes is "dogged," a significant word in this tale, by a stranger after he has been told of a terrible legendary hound on a remote moor who has seemingly caused a real death.[29]

> Sir Charles lay on his face, his arms out, his fingers dug into the ground, and his features convulsed with some strong emotion to such an extent that I could hardly have sworn to his identity. There was certainly no physical injury of any kind. But one false statement was made by Barrymore at the inquest. He said that there were no traces upon the ground round the body. He did not observe any. But I did—some little distance off, but fresh and clear.

"Footprints?"

"Footprints."

"A man's or a woman's?"

Dr. Mortimer looked strangely at us for an instant, and his voice
sank almost to a whisper as he answered.

"Mr. Holmes, they were the footprints of a gigantic hound!"[30]

Was Sir Charles the victim of a curse placed upon his family when
a fiendish ancestor, hunting with dogs, chased a girl to her death on
the moor for resisting his advances? That malefactor perished after being
hunted by a supernatural hound. His blameless descendant, Sir
Charles, is found dead. Holmes is summoned by the prospect of the
equally innocent heir arriving to take up his inheritance, quite
unsuspecting of its dark and Gothic shadow. To Dr. Watson's surprise,
he and Sir Henry are sent to Baskerville Hall, while Holmes stays in
London demanding detailed reports. Dr. Watson, determined to protect
his charge, is nonetheless impressed by the brooding moor and its many
natural dangers.

One fatal spot is Grimpen Mire, reputed to swallow ponies.
Frequent fogs obscure the narrow paths that are supposed to enable its
safe passage.[31] Possibly even more menacing is the danger from a vicious
escaped convict, while at night the wailing of an unseen animal evokes
the ghostly presence of the hell hound. Dr. Watson reports to Holmes
that one night, walking on the moor, he saw a mysterious figure who
is neither convict nor horrific hound.

The moon was low upon the right, and the jagged pinnacle of a
granite tor stood up against the lower curve of its silver disc. There,
outlined as black as an ebony statue on that shining background,
I saw the figure of a man upon the tor. Do not think that it was
a delusion, Holmes... He stood... as if he were brooding over
that enormous wilderness of peat and granite which lay before
him. He might have been the very spirit of that terrible place.[32]

In fact, danger to Sir Henry stems more from his enthusiastic
romance with Miss Stapleton. She is not what she appears to be: the
sister of neighbor and butterfly hunter, Mr. Stapleton. Holmes and
Watson deduce him to be the true villain. Also descended from the

Baskervilles, Stapleton is aiming to seize the family fortune by murderous means, and by forcing the cooperation of his so-called sister, who is actually his wife. For his plan, he has recruited the most dangerous dog he could find in London and painted him in glowing phosphorous, to simulate the legend. Merely the horrific appearance of the hound was enough to kill frail Sir Charles. By contrast, Sir Henry is determined to face down physical, erotic, and even spectral dangers.

While Sir Henry is more the traditional adventure hero than strangely absent Holmes, it falls to the great detective to save nature from the supernatural by uncovering Stapleton's plot. On the other hand, Holmes proves both more fallible and more identified with the irrational than is often assumed. When Selden, the escaped convict, falls prey to the hound, Holmes and Watson at first believe that they have lost Sir Henry, which would be a tremendous failure for the great sleuth. In fact, it is Selden who dies because he is wearing Sir Henry's clothes, given to him by his sister, who works at Baskerville Hall. The hound, trained to hunt down Sir Henry by the scent of a stolen shoe, correctly reads the clue of the clothing, while being unable to distinguish the man wearing it. Brute animal nature is more correct than human senses. The hound lunges for the right clothes while Watson's mysterious man on the Tor against the moon proves to be none other than Holmes himself.

Holmes is therefore fallible in failing to prevent a murder and also of a Gothic transgression of rational boundaries in his spooky appearance. Even more trickster-like is Holmes's paralleling with Stapleton, the villain. Each man "dogs" the other in London, exchanging places all too easily between hunter and hunted. Indeed, the trickster ambivalence between hero and killer becomes transformational in the dark underworld of the moor. Here Stapleton's revenge and greed is embodied by an animal whose nature has been abused into monstrous ferocity, while Holmes becomes a *genius loci,* a spirit of place whose nature as threat or guardian is unknowable right up until Watson draws his gun on an unseen figure.

The moor is the place where nature itself becomes Gothic; its traces of vanished people are signs of nature's dominion over man. Indeed, in Grimpen mire, the moor provides a metonym (the part connected to the whole it stands for) in its ability to eat whatever troubles its surface.

> Its tenacious grip plucked at our heels as we walked, and when
> we sank into it, it held us. It was as if some malignant hand was
> tugging us down into those obscene depths, so grim and
> purposeful was the clutch in which it held us.[33]

Holmes and Watson have no choice but to listen to Stapleton's horrible drowning in mud lest they suffer the same fate. To the moor, hero and villain are identical prey. It obeys no rational or generic rules of distinguishing good from evil, hunter from hunted, human from animal. The moor also refuses to offer the certainty of the villain's demise, given that no body is recovered. After such effacing of human cultural inscriptions, it is perhaps not surprising that, after being rescued, Mrs. Stapleton is not permitted to stay with the man she loves and has tried to protect. She too fades away while Holmes and Watson end in a London restaurant in an ironic taming of the moor's natural appetites.

Whereas modernity was supposed to save mankind from myth, Gothic reveals that it is not so easily banished. Indeed, myth returns to us in Freudian psychoanalysis's adherence to Oedipus, which is symptomatically a detective story in which the sleuth proves to be the murderer.[34] Although Holmes succeeds in exploding the myth of the Hound of the Baskervilles as a supernatural killer, like Oedipus, the revenge and sexual cruelty of the original tale recur in the present crimes.

So the myth of the hound is replaced by, or mutates into, the myth of the trickster in which not even the boundary between human and nature is preserved. Stapleton perverts the hound, and the moor triumphs over the villain, while almost swallowing the heroes too. Perhaps if trickster is developed into dismembered Dionysus we witness the dispersal of the divine spark into the matter of the devouring moor: Mother Nature in her most pre-Oedipal mode.[35]

It is time to look further at detecting myths of psyche and nature in our mysteries. For, from nature as Dionysian in "her" unreasonable appetites we will see emerging multiple, divine sparks of being—the goddesses.

From Dionysus to the God(desse)s: Detecting Psyche, Return of Myth

In the myth of Oedipus the King, Freud not only identified an early detective story, he also located it at the core of the modern psyche.

> The *Oedipus Rex* is a tragedy of fate; its tragic effect depends on the conflict between the all-powerful will of the gods and the vain efforts of human beings threatened with disaster... Modern authors have therefore sought to achieve a similar tragic effect by expressing the same conflict in stories of their own invention. But the playgoers have looked on unmoved at the unavailing efforts of guiltless men to avert the fulfillment of curse or oracle; the modern tragedies of destiny have failed of their effect... His fate moves us only because it might have been our own, because the oracle laid upon us before our birth the very curse which rested upon him.[36]

Oedipus, King of Thebes, seeks to save his city from a terrible plague. Rather than this being a medical problem, he is told that it is a moral and legal one. An unforgiveable crime has been left unrevealed and unrevenged, the murder of the former king, Laius, who was also the previous husband of Oedipus's present wife, Jocasta. Throughout the play, *Oedipus Rex*, by Sophocles, Oedipus acts as determined sleuth until the awful revelation. He was himself the unwitting killer of his own father, Laius, and has married his own mother, Jocasta. They even have children. With the suicide of Jocasta and the self-blinding of Oedipus, the play ends.

The truth of the crime does purge the city, yet at a terrible cost. Here the myth anticipates an underappreciated element of modern mysteries: what is revealed is often as traumatic, or even more so, than the original crime. Indeed, what the detective often discovers is that although an individual murder may be solved, it points to deeper crimes in human nature and culture that cannot be healed by a single hero of any kind. Freud decided that the haunting persistence of the Oedipus myth indicated that it spoke of a deep structure of the masculine psyche: that a boy child innately desires his mother and wishes to kill his father in order to gain her exclusive love.[37] The child will only let go of the maternal bond if he fears castration, the loss of genital pleasure that he realizes differentiates him from the mother's body.

Here the role of the psychoanalyst is to detect the clues that surface in later life by connecting them to this traumatic beginning. Somatic symptoms, malaise, and dream images are the traces of a primal wish to murder. Subjectivity begins in incestuous and murderous desire. Psychoanalysis detects the mystery of this founding story of consciousness, the myth-making of our being. Freud was less successful in accounting for how girls differentiate themselves from the mother's body, although he also detected a primal Eros for a parent of the opposite sex in the Electra complex. What remains important to mystery fiction from Freud's work are two striking possibilities: firstly, that myth structures psyche in ways not consciously known, secondly, the primal entanglement of love and murder.

In the fate of Oedipus, Freud saw the emergence of love for another as necessarily founded on the child just starting to realize that he is not totally part of the mother's body. Before the Oedipus myth has fully restructured the psyche, there is a pre-Oedipal mother to whom the infant is attached in a confusion of where he begins and ends. Hence the pre-Oedipal mother is not a woman, for the child as yet has no sense of bodily boundaries and differences between women and men. In order to distinguish genders, the child needs to know that his body is like the alien father's not the originating mother's. In this sense the pre-Oedipal mother is the ground of being, the whole universe that the child was once wholly absorbed into in the womb.

Beginning to sense the possible absence of the mother, the child struggles with desire mixed with fear of not having all his needs instantly fulfilled. Detecting a rival for the mother's exclusive attention, his longing for the mother is bound up with wishing to obliterate the father. Only after sensing the vulnerability of his penis and its difference from the mother will the boy child see himself as truly *detached*, different from the mother. He will begin to identify with the father as a masculine being. Later Freud came to see the drive to love and connection, and the urge to death and destruction, as innate and significant throughout life. Often termed Eros and Thanatos, after the gods of love and death, these drives are arguably crucial to the mystery genre.

In focusing on murder, mysteries do not only evoke Oedipal conflicts from a reader's archaic past in early childhood. Mysteries also map and enact a *playful* exercising and perhaps *exorcising* of Eros and Thanatos.[38] They allow these gods to breathe in our psyches in ways that are ultimately not too terrifying to bear. The playfulness comes from the way mystery fiction is so self-conscious and self-referential. As early as fictional Dr. Watson publishing stories about Sherlock Holmes in newspapers, mysteries have kept before the reader the notion that *this is a particular type of story that they are reading*. Mysteries refer obsessively to other mysteries, or find ways of reminding readers that this genre has rules and expectations.[39] Of course the fun is in how individual novels fulfill generic expectations in unpredicted ways.

So death is portrayed as real and painful, and yet on another level as fictional and *solvable*. Uniquely in mysteries death is a mystery that can be solved as long as the reader is simultaneously aware that this is fiction. The death drive is enacted and tamed. The taming is done by its entanglement with Eros, for this particular god of love is one of connection and bonding to life. At a deep level, the trickster genre itself structures such a bond with the reader, because the trickster is an image of survival against the odds. S/he lives to hunt or be hunted again.

Mysteries keep their promises to reveal the killer, even if justice, in the sense of assuaging all crimes in this world, proves unachievable. Therefore, Eros becomes as big a part of the reading experience as Thanatos in the recall of Oedipal fantasies. We bond with the detective and witness him at work by bonding with the characters and suspects. If the drive to death is blinding (as Oedipus was blind in the sense of being in the dark about what he was doing when he killed Laius), then the drive to life in mysteries is bound up with our need *to know*.

Oedipus, the successful detective, blinded himself because, true to the trickster basis of the genre, he was also the murderer being (self)punished. No longer blind to the truth, he perhaps anticipates how modern detectives sometime feel the need to blind themselves to crimes they cannot solve alone. *Oedipus Rex,* the play, is a tragedy that points to one of the core divisions in the modern mystery referred to above: the split between cozy and hardboiled. That this split is also gendered will be considered in Chapter 2. Here it is worth noting that both fictional types are visible in the Oedipus myth and in the tales of Sherlock Holmes.

The hardboiled mystery is linked to urban landscapes in which, like the original Thebes, there is a plague or dis-ease apparently inflicted by an unsolved murder. Unlike the Oedipus myth, discovering the murderer does not solve the problems of the city. Rather what is revealed is that urban decay defeats individual heroism whether it is revealed in the "tough guy" literally fighting with criminals or heroism in the form of the pursuit of knowledge. From these ancient Oedipal roots, the hardboiled arose in early twentieth-century America in the disillusioned works of Raymond Chandler and Dashiell Hammett.[40]

By contrast, cozy detective fiction began with the English contemporary figures of Agatha Christie and Dorothy L. Sayers.[41] Reeling from the loss of a generation of young men in World War I, the English looked for stories with largely rural settings in which the solution to the crime restored the small community to health. Cozies, as W.H. Auden memorably put it, aim to restore Eden, a garden of perfection only troubled by one serpent, the murderer.[42] While addressing Oedipal anxieties, the cozy embraces comedy rather than tragedy. The world is not irredeemably corrupt; the family, which is often the heart of the narrative, can be saved and Eros is often prominently displayed in romances between characters or with the detective. In the cozy, knowledge is ultimately healing, if not without trauma.

Another mythical way of looking at the persisting division between hardboiled and cozies is by way of the myth of the Holy Grail.[43] Here again is the notion that mysteries embrace myth as different types of knowing. In the legends, the Holy Grail is pursued because finding it will heal the Wasteland that is metonymically embodied in the sickness of its ruler, the Fisher King. In order to find the grail the purest knight, in the form of the indispensable integrity of the fictional detective, has to learn to ask the right question. In cozies, finding the truth, the Grail, heals the wasteland, but fails to do so in hardboiled mysteries, hence the tragedy that besets them.

What is significant to both types in this discussion of myth in mysteries is that with the Grail, Eden, and the Trickster, not to mention Dionysus, we are not relying solely on Oedipal mythical ways of knowing and being. Such a plurality of possible mythical expressions brings us to the work of C.G. Jung, who argued that Oedipus was only one of the many figures possible to psyche.[44] It is Jung who provides a framework for the different myths of goddesses that this book uncovers in modern mysteries by women. But first, another case history.

Case History (2): *Clouds of Witness* (1926) by Dorothy L. Sayers[45]

Clouds of Witness is the quintessential English cozy mystery taking place in the country house of a wealthy and aristocratic family. Gerald, Duke of Denver, brother to the amateur detective, Lord Peter Wimsey, is arrested for the murder of a guest, Captain Cathcart, whom he caught cheating at cards. Engaged to the Duke's sister, Lady Mary Wimsey, Cathcart is discovered shot dead and the Duke refuses to account for his whereabouts during the fatal moment. Lord Peter Wimsey, accompanied by faithful manservant, Bunter, arrives to find that the police investigation is headed by his friend, Inspector Parker of Scotland Yard.

At this point the Denver family does indeed represent a wasteland, in that all of them are frustrated and variously imperiled. With the death penalty in force for murder, Gerald's silence about his lack of alibi is potentially fatal to him as well as infuriating for his brother. Lady Mary is having a breakdown or behaving guiltily, as Lord Peter surmises. With Gerald's wife antagonistic and unhelpful, even the remarkable mother of these troubled siblings can offer no solution to what ails the family. True to the Grail myth's posited roots in fertility rituals, the mystery is bound up with sexuality that fails to channel a healing Eros. Mary, Gerald, and Cathcart are all involved with affairs that mis-lead them down the paths of knowing.

Cathcart is in a long relationship with a mistress who has drained him of money and finally deserted him. Cheating at cards is a desperate attempt to stave off financial ruin that proves instead a social disaster, since it is considered unforgivable in a "gentleman." Mary, having agreed to marry Cathcart in a fit of despair, is actually in love with a political agitator, and was running off to be with him when she hears a shot. Her very real distress stems from her mistaken belief that she has to choose between saving her brother or her lover. Finally Gerald has been conducting an affair with the beautiful wife of a local farmer, Grimethorpe, who is jealous and violent. Gerald's silence is both the code of a gentleman and has a real virtue, in that when the affair becomes known, the farmer's wife has to flee for her life.

As the detective in grail knight mode, Lord Peter can only discover the truth that will become the healing grail if he asks the right questions. In particular he needs to read the signs of Eros correctly in his siblings and the victim. Tellingly, he gets a vital clue after experiencing near death himself. In a suggestive echo of *The Hound of the Baskervilles* and its villain-swallowing Grimpen mire, Lord Peter falls into a bog and nearly drowns.

> "I tripped right into it," said Wimsey's voice steadily, out of the blackness. "One sinks very fast. You'd better not come near, or you'll go too. We'll yell a bit..."[46]

> For a few agonizing minutes two pairs of hands groped over the invisible slime. Then: "Keep yours still," said Bunter. He made a slow, circling movement. It was hard work keeping his face out of the mud. His hands slithered over the slobbery surface and suddenly closed on an arm.[47]

Here the love between the two men saves Peter because Bunter puts himself in danger to reach out to him. Both are then rescued by Mr. Grimethorpe, and it is in his bedroom that Peter discovers a letter to Gerald whose date will prove a substantial enough alibi for his brother's movements on the night of the death. Only by being close to death himself does Peter get access to this clue. He has to experience being buried alive, in a horrific repetition of his past trauma in World War One's trenches, to get him into the actual *place* of his brother, the bedroom, on that fateful night. Trickster-like Peter takes the path of the supposed murderer, Gerald, and has to enter the body of the earth to do so.

Such a literal entry to the underworld as under earth or mud proves critical to Peter correctly reading his own instant desire for Mrs. Grimethorpe as also embodying his brother. Oedipally, Gerald suffers for possessing the forbidden woman, whom he desires as his (m)other or that Other that psychically stands for her in his infantile Oedipal structures. The father, here in the form of the law, will take castration as far as the full expression of Thanatos in executing Gerald for a murder he only literally did not commit. Metaphorically of course, in taking unlawful possession of Mrs. Grimethorpe, Gerald has killed his father.

Lord Peter sees this Eros in cultural terms as the activation of legendary sexual privileges of his class. So, here the mythical dimension of detective fiction also serves as a vehicle for social criticism.

By himself entering the body of Mother Earth in a near loss of his life, Peter suffers the Thanatos as well as Eros aspect of the feminine and sexuality. The result is the discovery of writing (the letter) that enables him to ask the right questions of his "other," the brother who has acted on desires that Peter also feels. In the mystery genre, Trickster here plays with the characters of sleuth, supposed killer, and even the victim to suggest they are liminal, and not wholly separable. Cathcart also has been destroyed by Eros that has led him into the lap of Thanatos. In *Clouds of Witness* the healing grail and Eros, tuned to life and fertility in the sense of right relationships, is incarnated in family where sexuality is, for now, pushed into the margins.

While Mrs. Grimethorpe and her daughter escape from her husband, who represents Thanatos out of Eros as deadly as that afflicting Cathcart (who proves ultimately to have committed suicide), Lady Mary discovers she can trust her brother more than her lover. Detecting the truth enables the restoration of the Wimseys as a family who love and understand one another. Sibling Eros largely displaces sexual Eros, except for a promise of future fertility. This wasteland is healed to include Mr. Parker's declared feelings for Mary and her intimations of a possible future with him.

While the Oedipal myth, trickster, and the grail narrative do haunt this cozy mystery, the embodied process of knowing here suggests a need to expand the mythical terrain. Peter, with Bunter and the helpful Mr. Parker, tries the Holmesian methods of hunting for material clues and also the characteristically "cozy" method of talking to the involved group of suspects. Both prove frustrating in sequentially producing and promptly eliminating possible alternatives to Gerald as murderer. What changes everything is Peter's falling into a death-like grip of Mother Earth. As a result of where that initiation takes him, he finds himself in the literal place of the man he wants to save, learning to read his own sexual desire as a knowing of the mystery. Might Aphrodite also be figuring the search for a healing truth in *Clouds of Witness*?

Perhaps the Freudian Eros and Thanatos mutuality could be read better here as Aphrodite emerging from an archaic pantheon of the sacred Earth herself? Can Aphrodite be a way of knowing as well as a structure of being? Here we can recognize Freudian Eros and Thanatos as entangled in a Dionysian frenzy that mixes sexuality with death. From this step into a bigger pantheon comes the movement of Jung's twofold Dionysian dismemberment: first "opposites" of Eros and Thanatos, then their more rejuvenating dispersal into the matter of many god(dess)es. In this way mysteries require a larger notion of myth and psyche than that permitted by Freud.

Jung and Goddesses: Stories of Knowing and Being

C.G. Jung broke off his close association with Sigmund Freud because he could not accept that sexuality was the fundamental mode of psychic energy and so making the story of Oedipus the sole founding myth of consciousness.[48] He wanted more myths. Jung also considered that the psyche looked forward to future developments, and therefore would seek out and inhabit a number of mythical narratives in consciousness. At base, Jung took very seriously what I like to call his founding principle.

> Nobody drew the conclusion that if the subject of knowledge, the psyche, were in fact a veiled form of existence not immediately accessible to consciousness, then all our knowledge must be incomplete, and moreover to a degree that we cannot determine.[49]

To Jung the existence of the unconscious, that part of the psyche that we do not control and which produces dreams, means that all knowledge must be regarded as provisional. We cannot be sure that we know anything as absolute truth, for we do not know what we do not know. Such ignorance should be regarded as a challenge to what we think we know. In this context, "myth" acquires a special charge for being the narrative structures of the psyche. Myths are psychologically the stories that transform what we think we know into stories of who we are. Put in terms of Jungian psychology, myths define and shape relations with the unconscious.[50]

For Jung, taking the unconscious seriously meant regarding theory as a pragmatic and provisional attempt to work with the psyche, not some route to absolute truth. Archetypes were his name for those inherited unconscious properties that tend to structure the same sorts of images and meanings throughout human cultures.[51] Archetypes are only the *potentials* for signifying for they require, and are in-formed by the history and society from which they emerge. Here myths are the stories charged with archetypal unconscious energy that mold consciousness. Myths are the narratives generated by the meaning-fertile archetypes in conjunction with the life of the psyche as embodied in time.

Archetypal images are unions of body, history, culture, and a psyche that cannot be determined by these entities. They are imbued with myth because of their tendency to accrue meaning by producing narrative. Here Jung, a contemporary of modernist artists, offers a psychological counterpart to their excavation of myth as a dynamic and living aspect of who we are. His notion of myth also looks back to that of the Ancient Greeks such as Plato and Aristotle. As described by Laurence Coupe in his excellent book, *Myth* (1997; 2009), both philosophers regarded myth as the fundamental ground of knowledge, while differing about the importance of narrative itself to knowing.[52]

Plato argued in favor of *Logos* in the sense of an abstract kind of truth that could be considered as separate from the myth generating it. By contrast, Aristotle argued of history that *Mythos*, plot or story, could not be separated from the knowing. The story in history is what we know and how we know it.[53] Mythos becomes an irreducibly narrative way to truth. In terms of mysteries, logos differs from mythos in taking every deliberate act of murder as the same kind of truth, in that there is a prohibition against killing that is absolute and transcendent of actual circumstances. Alternatively, each murder is unique, is mythos because it is immanent to the story of the crime.

Evidently there is a need to connect mythos and logos so that justice is not either wholly relative or wholly blind to the specific conditions and motivations of the crimes. *Dialogism* was devised by language critic M.M. Bakhtin in the early twentieth century to describe something that only appears in a dialogue between two or more interactions.[54] In this sense, mysteries are a dialogic form, in that they are narratives in which an accepted logos—that murder is always forbidden—is in dialogue with a mythos showing how and why such a crime is committed, usually by someone who is not an habitual criminal.

Dialogism also works for Jung's archetypal images in that they are made from a dialogue of actual historical circumstances and the unconscious, meaning-seeking energy in the archetype.[55] Hence, myth for Jung is dialogical, a narrative produced by the stories that are embodied in the world (mythos) and the unconscious psyche, undermining the absolute nature (logos) of any secure knowledge posited by myth.

Here Jung's difference from Freud also becomes apparent. So secure was Freud in the Oedipus drama as the founding structure of consciousness that the Oedipus complex (of son trying to remove his father to keep his mother to himself) is more logos than mythos. By contrast, Jung's adherence to the properties of the unconscious psyche as incalculable by consciousness shows that his narratives of consciousness are plural and remain embedded in mythos.

Of course, Jungian psychology offers stable concepts, in notions such as archetypes and individuation, which is a continuous mythical and fluid restructuring of a person's psyche. However, Jung was clear that concepts are not to become pure logos; they are not to be detached from the mysteries of psyche. His concepts, because they require clarity and rationality, will not account for the deepest and most alive aspects of the psyche.

> We have to break down life and events, which are self-contained processes, into meanings, images, concepts, well knowing that in doing so we are getting further away from the living mystery. As long as we ourselves are caught up in the process of creation, we neither see nor understand; indeed we ought not to understand, for nothing is more injurious to immediate experience than cognition. But for the purposes of cognitive understanding we must detach ourselves from the creative process and look at it from the outside... In this way we meet the demands of science.[56]

Here, Jung's notion of "living mystery" includes the larger sense of "mystery" identified in this book as crucial to detective fiction. Mystery points to both that there is a crime to solve, a whodunit, and, that complete and absolute truth about the crime, like absolute knowledge itself in Jungian terms, is impossible. There is always more mystery in mysteries. It therefore follows that mysteries cannot promise complete and absolute justice, for without perfect knowledge there can be no perfect restitution.

Mysteries always dream of justice, while offering something more fallible and partial. On the other hand, yet another aspect of the dialogism of its trickster myth is that the detective embodies for the reader the desire for logos, or perfect truth, even as she or he discovers that mythos can never, perhaps should never, be banished. No one suffers more the anguish of imperfect justice than Sara Paretsky's impassioned crusader, V. I. Warshawki, as we will see in the next case history.

Case History (3): *Deadlock* (1984) by Sara Paretsky[57]

Private Investigator, V.I. Warshawski, did not mean to embark on a quest with regard to the unexplained death of her cousin, Boom Boom. Rather, guilt that she has not sufficiently nurtured the ex-hockey player turned grain importer, and the mystery of why he left an urgent message for her, impels her to question the assumption of an accident. Boom Boom fell off a wet dock near the offices of Eudora Grain while on his way to talk to someone at Pole Star Shipping. He said that he had information that recent vandalism at Pole Star was a bigger problem than previously assumed.

Plunged into Lake Michigan, Boom Boom was horribly mangled by a ship's propellers. When Boom Boom's apartment is burgled, V.I. becomes convinced of something more sinister than an accident is afoot. In addition, during the burglary, the building supervisor is killed while protecting the apartment, at least partly at V.I.'s request. Nothing rings true to V.I. about Boom Boom's death, including the reaction of his supposed girlfriend, ballet dancer, Paige Carrington, whose elegant lifestyle has no visible means of support.

What distinguishes *Deadlock* and Paretsky's work as a whole is how intimate and family relationships become entangled with troubled social and industrial networks. Great Lakes shipping companies prove driven by personal obsessions as well as by the economic structure that encourages greed and criminality. Niels Grafalk, who inherited vast wealth and several industries, is obsessed with the idea that his shipping company represents his Nordic patrimony. He bitterly resents the defection of a protégée, Martin Bledsoe, who recently began Pole Star Shipping and became a keen, if small, competitor. Small acts of willful damage to Pole Star vessels culminate in a bomb detonated below the waterline of Bledsoe's major ship.

> We were rising again. We didn't have the buoyancy of a ship in water, but rocked as if balanced on the air itself... I didn't understand what was happening, why we were rising, why there was no water pushing us up, but I felt vilely sick.[58]

> ... "It was a bomb, you know. Depth charges. Must have been planted right on the center beam. Set off by radio signal. But why?"[59]

In *Deadlock* murder is a violation of natural forces as well as an extension of human greed and obsession. A ship launches into the air, losing contact with its watery home. Four crew members die in the blast, including a man who drowns in the barley cargo, suffocated by the fertile grain. And an athlete who won fame on the ice sinks underwater.

V.I. herself is drawn to stir the depths that everyone around Boom Boom would prefer unstirred. This includes the police, who are keen to classify his death as accident. They are unsympathetic to V.I. since respected officer, Bobby Mallory, wishes her to remain in the domestic role he considers proper for women. As the police do not want to come up against the vested interests of the rich, they ignore V.I.'s instincts by treating the building supervisor's death as unconnected to that of Boom Boom.

V.I. is a crusader for justice who half-jokingly calls herself a female version of Don Quixote, one who is radically at odds with the modern corrupt world.[60] As a hero in an unheroic age, V.I. risks doing harm in her pursuit to do good. Like the bomb in the depths of the ship, V.I. penetrates the depths and sets off deadly energies. The building supervisor's widow accuses her of getting her husband killed by pointing him towards the fatal apartment. An attempt to kill V.I. by sabotaging her car results in several fatalities, leaving her injured and more determined.

While in his detachment Sherlock Holmes fails to prevent a murder on the moor, arguably V.I.'s very personal, emotional involvement precipitates more killing. In this she takes on the pain and suffering of being both trickster-destroyer and victim of the genre's trickster myth. The web of deceit, passion, and greed snuffs out Boom Boom trickily because the manner of his death conceals the actual murder. Aiming to assuage her sense of family guilt and get justice for Boom Boom, V.I. unintentionally inspires further deaths.

Like Dorothy L. Sayers's Lord Peter Wimsey in her attraction to Martin Bledsoe, V.I. finds sexuality part of the sleuth's quest. They have one date before he becomes a possible suspect for her attempted murder. V.I. decides to evaluate him personally by visiting him on his ship, and therefore is present when the bomb destroys it from below. *Deadlock* offers a watery underworld in which what enables life becomes deadly. Human ingenuity is perverted to irrational and greedy designs. Entering the business environment of grain importing and shipping, V.I. discovers that the basics of life, bread, and water can constitute an underworld in which murder stands for the mystery of a corrupt city within wider society.

Boom Boom's death is revealed as a metonym, as it stands for a larger narrative of crime. In turn, criminal acts are generated by rage, acquisitiveness, and the destructive nature of a class divided society. Unable to psychically bear an impoverished background connoted with being despised, Bledsoe, Paige Carrington, and others are led into crimes that provoke even darker shadows.

Above all, V.I. is physically and psychically implicated in her mystery. Her love for family drives her to greater involvement because *she is already involved*. Driven by Eros, love, and connection leading to knowing, she makes use of her sexuality without exploiting it or the man with whom she feels this connection. V.I.'s fierce drive for justice transmits into a sense of purity of life that has its underworld aspect in also precipitating death (shown to be characteristic of Artemis in the next chapter). Like Holmes, she rescues nature from supernature by solving the perversion of what ships do and grain provides. More like Peter Wimsey than Sherlock Holmes, she bears the marks of her detecting on her body, and as a consequence is almost swallowed by the elements (for her, water; for Wimsey, drowning in mud) of earth itself.

Hence V.I., like Lord Peter, models knowledge through bodily knowing and connection, rather than disembodied separation. In the next section of this chapter, these two types of knowing will reveal a mythical heritage.

The God and Goddess of Creation Myths

Ultimately, another dialogism haunts both Jungian psychology and mysteries in their mutual entanglements in myth. Ann Baring and Jules

Cashford in *The Myth of the Goddess* (1991), suggest that modernity has been created by the intersection of two types of creation myth that serve as two, mutually defining, structures of consciousness.[61] Such persistent dualism stems from mythical transitions in prehistory. Firstly, early religion regarded the earth as sacred, mother to all life, to women and men as born equally from her womb along with all we now call nature. Consciousness, therefore, was based on connection, to nature, to animals, to the other gender, to the divine. Sexuality and the body were also ways to the sacred, for connection was the path to the numinous.

Sacred Earth beget animism, the sense of a divine multiplicity because all matter was in-spirited. Historically, it seems that animist Earth Mother religions were succeeded by monotheism, which, by providing a disembodied God who created nature and everything as separate from himself, founds dualism. As a result, the dominant monotheism modeled consciousness is based on separation and difference. When culturally structured as patriarchy, rule of the father drawn from a Father God who is omnipotent and bodiless, it meant that women, sexuality, body, connection, and non-human nature all became "other" in an inferior sense.

Baring and Cashford emphasize that neither creation myth is inherently superior. Rather, human consciousness to be strong and viable needs both a sense of differentiation from the "other" (not least in becoming a separate psyche in early infancy) and connection to the other. In Western modernity such connection means re-connecting to one's own body, for patriarchy has modeled identity on divorce between psyche and body through God's separation from nature.

The problem, suggests *The Myth of the Goddess*, is the imbalance between these mutually defining creation myths. In actuality, domination by Sky Father separation has hardened into a patriarchy that has devastated the planet and erected untenable power structures between people. Healthy consciousness is dialogical between both types of creation myth; the psyche needs co-creation through Sky Father separation mitigating Earth Mother connection. Put another way, the psyche needs more than one myth, which also means that it needs more than one myth of knowing.

Sky Father is the engine of logos, of knowledge as detachable and *separate* from story, which here also means detectable from the body and the psyche as experienced as storied mythically. Earth Mother sponsors mythos, knowledge as inseparable from story, from myth meaning here the psyche as embodied and connected to the other as unconscious, other people and the being(s) of nature.

What this means for mysteries is that they also exist dialogically between mythos and logos, Earth Mother and Sky Father. Each mystery has its own story, its own mythos, and is also understood as a particular type of story; a distinct genre whose logos is the mystery form. What is additionally becoming apparent is that the relentless duality haunting mysteries needs to give way to something more plural and flexible, in which unique texts can manifest in relation to recurring myths in the psyche, culture, and history.

In the next case history, we will see the emergence of two goddesses who are not pushing their hero into singular Oedipal consciousness. Nevada Barr's sleuth comes to know, even to be, the goddesses Artemis and Aphrodite in ways considered here and in the following chapter.

Case History (4): *Ill Wind* (1995) by Nevada Barr[62]

While in *The Hound of the Baskervilles* the key drama takes place in a dark underworld of the moor, where prehistoric huts show traces of a vanished people of which nothing is known, Nevada Barr's *Ill Wind* is set in Mesa Verde National Park in Colorado, at the site of the vanished and enigmatic Anasazi native people. The novel opens with the mystery of this sophisticated culture that suddenly dispersed eight hundred years ago, a mystery not solved within the story, while still haunting it. Here mysteries evoke their Gothic ancestry in a past that resists modern modes of knowing.

Anna Pigeon, widow and National Park Ranger, is drinking too much and barely managing to cope with her job of law enforcement in a community of fellow rangers, tourists, and the interpreters of the Anasazi ruins hired to serve them. When Stacy, a ranger in a difficult marriage with a disabled child, is found dead in the historic site, Anna wonders if there is a connection with domestic disputes and machine sabotage already taking place. Her investigation is hampered by her own erotic vulnerability to Stacy. Falsely told that Anna was sleeping with her husband, Rose, the widow, refuses to cooperate and prefers the male FBI agent, Stanton, sent in to take over the case.

Stanton is wise enough to value Anna's knowledge of the community. The two of them talk to Jamie, a self-dramatizing young woman convinced that spirits of the Anasazi, or Old Ones, remain on site to punish those who violate its secrets. Her emotional convictions lead Anna to witness a ghostly veil on the night of the murder, and eventually to the discovery of illegal toxic waste being dumped in the park. The ghostly veil is indeed deadly, but is a luminescent manifestation of poison from modern life rather than revenge of the Anasazi. Without the compassion and imagination to listen to Jamie, Anna might not have uncovered this crime and solved Stacy's death.

One of Anna's difficulties in the Mesa Verde is lack of solitude. Even the night is uncomfortable to her because this ground is so intensely cultivated by human presence; even if, unlike the New York City of her past, its darkness is filled with more dead people than living.

> Walking the trail in Texas, skirting islands around Isle Royale, she'd worn the night like a star-studded cloak. But Mesa Verde was all about dead people. In the mind – or the collective unconscious – there was a feeling they'd not all left in the twelve hundreds… Dreams and desires haunted the mesa the way they haunted the rooms in old houses.[63]

In fact, Anna's sensitive attunement to the natural world pays off in two ways as a detective. Comfortable and stealthy in nature, she comes upon strange human practices such as Bella, a child with dwarfism, playing a game of pretending to be dead. Uneasy in the psychically peopled Mesa Verde ruins, her senses are alert to minute traces of crimes, and are also imaginatively able to sense the tones of others' feelings in their passion for this land.

Yet Anna is no misanthropist. Lonely for female companions, she is drawn into a friendship with Bella and those caring for the child. Like Aphrodite, she enjoys sharing beauty, such as when she helps visitors to see the grace of Anasazi houses and artifacts.

> Sharing beauty with total strangers made the world seem a friendlier place. In a culture dominated, if not by violence, then certainly by overheated reports of it dished out by a ratings-starved news media, it reassured her that the love of peace and natural order was still extant in the human soul.[64]

In the next chapter we will consider significant elements of Anna's mode of detection, in her need for solitude and nature, her role as protector of children, sponsor of the experience of beauty, and her sensitivity to the borderlands of life and death. We will see how Anna's distinctive ways of knowing and being amount to re-embodying, through the protean energy of the trickster genre, the goddesses, Artemis and Aphrodite. If the trickster is the shape-shifter of both genre and sleuth, s/he is also Dionysus dismembered into multiple divine possibilities. These are then enlivened, resurrected, renewed, and reborn as goddesses in women authored mysteries. I will show in Chapter 2 how through the divine-haunted detective the goddesses are invoked in the psyche of today's readers of mysteries.

CHAPTER TWO

THE GODDESSES FOR WOMEN WRITERS
GENDERING THE GENRE

Introduction

I n Chapter 1 I showed that the detective or mystery genre is haunted by a dualism straining to explode into multiplicity and diversity. Whether that dualism is considered in terms of Western modernity's dominant monotheism (in which a transcendent god structures nature as other), in Bakhtin's dialogism, or in Hillman's discussion of the opposition of Apollo and Dionysus, I suggested that the most comprehensive dualism is in the positing of two types of creation myths of being.[1] Sky Father and Earth Mother are hypothetical form-making narratives in which consciousness is either based upon separation into dualism, or, alternatively, founded on connection and embodiment.[2] They represent two, mutually implicating tendencies in psyche and culture, not actual myths, not least because they are too stark to work alone. Humans require more than just either one. In fact, a viable psyche and culture needs these tendencies in a creative dialogue.

So consciousness requires the presence of *both* a Sky Father and an Earth Mother. A Sky Father is unavoidable, because without separation from the mother in such fateful narratives as the Oedipus complex, no new identity is possible. On the other hand, without an Earth Mother psyche through body and connection there is neither love nor reproductive desire. Baring and Cashford's *The Myth of the Goddess* persuasively argues that Western modernity is sick because of its excessive repression of Earth Mother and neurotic over-favoring of Sky Father. Although the depth psychologies of Freud and Jung bring back an Earth Mother as the pre-Oedipal mother, this brooding figure prior to gender differentiation has not yet succeeded in rebalancing the modern psyche.

One crucial aspect to Earth Mother and Sky Father is that these myths do not constitute two genders based unproblematically upon bodily sex. Put another way, an Earth Mother and a Sky Father do not imply essentialism, the notion that persons with female bodies have an innate gender identity as feminine and vice versa for men. The idea of an unchanging psychic gender from bodily sex has largely been discounted through pioneering research into varying styles of gender across different cultures and throughout history.[3] On the other hand, bodily sex is implicated in gendered behavior. Moreover, in championing an embodied psyche where psyche and soma are inextricably bonded, depth psychology inevitably encompasses bodily and sexual differences.

To return to the creation myths and their non-essentialism, Earth Mother is the pre-gendered pre-Oedipal *ground* of being, because mythically she is the divine planet who gives birth to all life; while psychologically she is the mother from whose body the infant emerges. In this sense, the mother's body can be an actual woman, or any artificial gestation chamber, for to the infant no human shape is yet discernible. Hence the Earth Mother may be a goddess, but s/he is not a woman, although the feminine pronoun will be used in this book for convenience. An Earth Mother is a goddess of both women and men, just as the planet nurtures and produces both, or all, possible genders.

By contrast, a Sky Father takes an active and differentiating role in gender creation. By making nature and the planet as separate from himself, "he" founds duality as a fundamental condition of being. Once that duality is inscribed onto the different bodies of women and men, gender as a set of binary alternatives begins. Furthermore once gender differences are mapped onto the relation between natural and supernatural, then the divine becomes masculinized and "his" creation feminized. Therefore, in the gendered *position* of God a model of consciousness emerges as disembodied and masculine. It takes only a cultural literalization of God as a disembodied source of fertility to erect patriarchy, the rule of a father who is patterned on the divine Father of created, dumb (in both senses) nature.

Psychologically, the Sky Father is the Oedipal father whose intimated threat of castration separates the child from amorphous pre-Oedipal mother and forces recognition of two gendered bodies. Sky

Father makes two genders and in so doing becomes one of them. In this way, Earth Mother and Sky Father are not so much woman and man as holism (everything connected) and dualism (everything defined by an-other). The human psyche needs both myths as basic structures of consciousness. Even in the most patriarchal environments, meaningful connection, body, and their psychic energization as Eros remains dynamically alive. Without some Sky Father elements, there would be no consciousness, for the psyche would be a boundless space of multiple, chaotic impulses.

Consequently, the alternative to dominance by dualism in patriarchy is not switching to a simple matriarchy, of rule by mothers, nor is it to jump into Earth Mother myth of no differentiation at all. Switching the gender in charge, or changing one framing myth to the other, is to remain trapped in the same dualistic structure and become coercive by its privileged status. Rather, healing the world is a matter of rebalancing both creation myths as necessary and necessarily inter-related. Only when both myths are creatively entwined can they be fertile in sponsoring the fullness of our possibilities.[4]

Given that Earth Mother invokes animism, many spirits in nature, the psyche requires not two divine beings but many. Put psychologically, as James Hillman proposed, the psyche needs polytheism, *many* goddesses and gods of being and knowing.[5] For these divine figures offer gender as polymorphous and protean in our possession of, or possession by, many divine stories.

That these goddesses and gods are modes of knowing is an aspect of their function in structuring consciousness. For example, in the previous chapter, *Clouds of Witness* found Aphrodite in the sleuth's recognition of the erotic current of the crime through his own sexual attraction to his brother's mistress. Artemis infused Anna Pigeon in her quest to know through her liminal presence to non-human nature, while V.I. Warshawski is both Athena guarding her city and Hestia trying to reconnect the family that embodies her hearth fires. How these goddesses stem from the trickster myth of the mystery genre and Dionysian dismemberment will be considered next.

The drive to diversity in contemporary mysteries has previously been explored by Stephen Knight in his comprehensive *Crime Fiction 1800-2000: Detection, Death, Diversity* (2004).[6] While covering a longer span of time and authors of both genders, Knight's invaluable analysis stops short of considering myth as a productive factor. He prefers to see history and power as determining forces, and so his work differs markedly from my own archetypal approach.

It is time to look at how genre operates dynamically for the goddesses, from trickster myth to divine vehicle.

Trickster, Divine Dismemberment, and Goddesses

Jung shows the way from the trickster myth of hunter/gatherers to the return of the Earth Mother in order to heal the modern psyche.

> Even [the trickster's] sex is optional despite its phallic qualities: he can turn himself into a woman and bear children... This is a reference to his original nature as a Creator, for the world is made from the body of a god.[7]

Here in Jung's work is the admission that the trickster is one image or figuration (figure) of the Earth Mother; his more capacious sense of the Freudian pre-Oedipal mother. For the trickster is the protean psyche itself, as the origin, ground, and multiple possibilities of our being. The trickster as pre-Oedipal entity makes a cosmos from her/his ungendered body. Lewis Hyde, in *Trickster Makes this World*, hypothesizes that the cross cultural trickster myth stems from a deep archaic time when it became the psychic engine of humans hunting and being hunted.[8]

> To my mind, then, the [trickster] myth contains a story about the incremental creation of an intelligence about hunting.[9]

In my book, *The Ecocritical Psyche* (2012), I argued that mystery fiction activates the trickster aspect of psyche that has the potential to re-orient our relation to human and non-human nature.[10] This discussion will be developed through later chapters of this book, where I will elaborate on how goddess myths offer particular psychic interventions into nature and human nature through detective fiction. For now, I want to continue the theme of this chapter which is the conversion, within mysteries, of gender dualism into diverse styles of the feminine. As we have seen, such dualism is expressed mythically in the mutual relating of Earth Mother and Sky Father. Here the latter posits an originating gendered creator (masculine), and an "other" (feminine) as its negative; while the former, a pre-Oedipal mother, or trickster, offers plural and mutable gender potentials.

Such dualism is challenged by the notion of divine dismemberment, examined in Chapter 1, in which oppositions morph into a relational multiplicity in the mythical structure of Jung's rendering of Dionysus.[11] Ultimately, Dionysian multiplicity, which is also the matter (or mater) of trickster, takes on being in the different goddess and god myths, as I shall show below. In turn, the Dionysian sparks find soma and psyche in our writing and reading mysteries. Therefore dualism within mystery fiction, begun in gendered form, is mobilized by the genre's innate trickster-Dionysian energy into a creative expression and reinvention of gender styles.

Such a complex mythical history accounts for the perception of gender dualism within the early decades of the literary form. For the early twentieth century there appeared to be a real division between tough guy heroes and genteel feminine amateurs with appropriate gender assignments. Yet more recent female writers in particular, I suggest, have been busy dismembering such stark duality that reeks of gender stereotyping. Male hardboiled mysteries are no longer balanced against the female clue-puzzle, or "cozy," as the genre celebrates its own polytheistic liberation.

Gender and Genre: The Hardboiled Woman and the Cozy Man

By the twenty-first century, sleuth fiction by women in the United States and Britain disrupts the earlier gendered opposition of hardboiled and cozy, and also dismembers dualism into multiplicity in a self-conscious irony native to this trickster genre.[12] Mysteries continue to rely upon a sophisticated readership, one that recognizes the masculine and feminine roots of the hardboiled and the cozy. So, in order to scrutinize the undoing of gender opposition, it is worth looking at how the hardboiled and the cozy emerged in the early twentieth century mysteries.

Unsurprisingly, it was male writers of detective fiction set in American cities beset with organized crime who gave us the hardboiled hero. Raymond Chandler was not only prime practitioner; he also memorably analyzed his art of the iconic "mean streets."[13]

...[D]own these mean streets a man must go who is not himself
mean, who is neither tarnished nor afraid. He is the hero; he is
everything. He must be a complete man and a common man
and yet an unusual man. He must be, to use a rather weathered
phrase, a man of honor—by instinct, by inevitability, without
thought of it, and certainly without saying it. He must be the
best man in his world and a good enough man for any world.[14]

Chandler's hero, Philip Marlowe, is emphatically masculine, and
in fact seems designed to resist the trickster elements in both fictional
sleuth and his genre. In a sense Chandler is here archaizing the
hardboiled, in pressing the hero back into the ideal grail knight and
corralling the trickster energy into the deceits of the city revealed to
be an irredeemable wasteland. Here the detective as pure knight ends
his quest in becoming the wounded Fisher King, metonym for the
unhealed world and bearing the scars of its violence.[15]

By contrast, the clue-puzzle, or cozy, is dedicated to the success of
the grail quest, or the restoration of Eden.

The phantasy, then, which the detective story addict indulges is
the phantasy of being restored to the Garden of Eden, to a state
of innocence, where he may know love as love and not as the
law. The driving force behind this daydream is the feeling of
guilt, the cause of which is unknown to the dreamer. The
phantasy of escape is the same, whether one explains the guilt in
Christian, Freudian, or any other terms.[16]

Here is the coziness haunted by the unease of the cozy. Represented
typically in the women writers of an early twentieth century England
reeling from the wholesale male slaughter of World War One, the "clue-
puzzle," a term developed by Stephen Knight, was evolved by writers
such as Agatha Christie, Dorothy L. Sayers, Ngaio Marsh, and Margery
Allingham.[17] These women pioneered a form to exploit trickster
changeability that would enable a fictional escape from the all too literal
wastelands of the battlefields.

With the gender opposition of hardboiled and cozy, therefore,
comes also a cultural division between America, a country finding a
voice for its new urban realities, and England, struggling with an absence
of a generation of young men. Together with the gender division in
society between male soldiers and also a largely male police force, it is

easy to see a reciprocal influence between social role and the emerging of gender dichotomy in the two related subgenres.

And yet the trickster-Dionysian energy in the whole genre refuses to be channeled into a static dualism of masculine hardboiled and feminine clue-puzzle or cozy. In the later twentieth century, especially in the work of women writers, there is a skillful dismembering of stereotypes and the liberation of goddesses in mystery fiction. Moreover, a polytheistic approach to understanding the genre also demonstrates how some novels themselves diagnose the absence of a much needed goddess. For example, here I will suggest how Aphrodite emerges from, and also overshadows, Dionysus. To demonstrate, I will show how Marcia Muller's *Trophies and Dead Things* (1990) shows the high cost of banishing Aphrodite and shifting war from an Ares, her lover, to Apollo, a god without her mitigating embrace.[18]

Reclaiming Aphrodite

As Ginette Paris shows in *Pagan Meditations,* Aphrodite as feminine sexual energy and knowing was marginalized by Plato and much later, by Freud.[19] Each reconfigures Eros to occlude her radiance. In the Greek pantheon, Aphrodite shares with Dionysus in giving "a central position to the spontaneity of the body and to sexuality."[20] There the resemblance between more primordial and brutal Dionysus and civilizing Aphrodite ends. While Dionysus is ecstatic and savage, Aphrodite is beauty and seduction. Aphrodite is the arts of love while Dionysus provides instinctual energy frequently careless of human vulnerability.

Consequently, in the notion of dualism giving way to multiplicity via dismembered Dionysus, Aphrodite represents a further stage in the gods developing sophisticated arts and graces. Aphrodite is the civilizing force arising from the embodied psyche in sexuality. She is the capacity for beauty that inspires corporeal love. Paris sees Aphrodite as feminine sexual energy that bestows and responds to beauty beyond the nature culture divide.[21] Aphrodite is found in the seductive beauty of flowers, of aesthetic nature in gardens as well as in the gorgeous clothing of those seeking or expressing love. Not least of Aphrodite's cultivating capacities, then, is to enable nature and art to embrace in human enjoyment of the body.

Evidently, Aphrodite can then be seen as one of the Earth Mother's multiple figurations as divine earth as source of being and meaning. Or, she may be a reconceived version concealed in plain sight in a pantheon dominated by Zeus, a patriarchal sky father. To the latter interpretation falls her birth legend as arising from sea foam resulting from Cronos's castration of his Sky Father, the Sky himself, Ouranos.[22] Aphrodite then becomes a fluid, sinuous sexuality that reminds us that our bodies are mostly salt water.[23]

Above all, Aphrodite is a virgin goddess; her divinity demonstrates sexuality as sacred. Put another way, Aphrodite is a form of knowing and being in mysteries.[24] She is virgin in the sense of being self-contained, not to be regarded as defined by any one relationship to an-other. Significantly, she shares this virginity with Athena, Artemis, and Hestia, suggesting the importance of the feminine as a primary reality, and not as a secondary product of an original masculinity.

On the other hand, the project of marginalizing the sexuality of divine feminine began early with Platonic philosophy. As Paris explains.

> Platonic philosophy marked the end of Aphrodite's predominance; it gave the myth of Eros precedence of that of Aphrodite, dissociated love from its corporeal aspect, and valued mostly the all-male relationships… More and more Apollo controlled Dionysus… Woman's body stopped being one of the paths to the sacred.[25]

Here Paris reveals Plato's move in replacing Earth Mother as origin with Sky Father's dualistic and patriarchal values. By elevating mobile Eros and discounting the somatic sacred of Aphrodite's erotic love, his philosophy looks towards a divine devotion directly opposed to the pleasures of the body. Such a move is indivisible from a separation and privileging relation between masculine and feminine. Moreover, Apollo, god of reason and order, starts to direct cultural preferences, rather than being a psychic capacity equal to those possessed by very different gods.

In making Eros the *masculine* force of psychic energy that Freud saw as fundamentally sexual, Freudian psychoanalysis compounds Platonic preferences.[26] Hence the Freudian libido is masculine; Aphrodite and feminine sexuality as an ontological or primary reality are not offered as part of the Freudian tradition. This is clearly not the case for Ginette Paris's development of a Jungian notion of multiple archetypes, inborn psychic potentials all capable of feminine or masculine expression. Paris contributes magnificently to Archetypal Psychology by positing a polytheistic psyche in which goddesses exist as differentiated psychic powers.

> The masculine and feminine universes are in constant attraction and repulsion, interdependent and organically linked. We are here in the world of Aphrodite, Hermes, and, in general, of Greek polytheism: constant negotiations, many rivalries, but also great intensity of life.[27]

Such a portrayal is not an invitation to retreat from the present world. It is rather a way of re-connecting with human potentials that have been neglected in the Western psyche. If, as I have been suggesting, these goddesses and gods have returned in cultural forms such as mystery fiction, it might be worth exploring a little further what Aphrodite holds out to us. After all, if divine feminine sexuality is a civilizing force, then what about the sexual component of fictional crime? Might such an understanding of the knowing potential of the body aid a detective's quest?

Aphrodite is beauty in a state of grace, says Paris. She is a divine invitation to sexuality that is far from a facile endorsement of society's often perverse preferences for beauty in women and men.[28] Aphrodite enfolds the whole body in a glow that incites desire. She brings couples together without preference for heterosexuality. Her arts are culture-*making*, refining intense and beautiful human connections. She civilizes in ways that promote the liveliness of the body as a way to the sacred.[29]

Above all for this book, Aphrodite *knows* through her sexuality, her charged body, her connection and arts of love, her passions. Her taste for new relationships forms intimacies that reveal what would otherwise remain hidden to an investigator. Her embodied sense of being and touching activates a somatic, only partly conscious awareness that contributes to other, more rational and intellectual modes of thought. Sexually inspired, Aphrodite is beauty that communicates possibilities beyond conventional signals, offering more knowledge of relationships than more polite non-sexual interaction permits.

Indeed, Aphrodite is a possession of an enhanced capacity of what Michael Polyani calls "tacit knowing," comprehension that is embodied and intuitive.[30] We know without knowing *how* we know. Tacit knowing is creaturely skillful intimation that comes from our somatic roots in nature. So here, in her civilizing arts, Aphrodite provides tacit knowing through instinctual sexuality that is also psychically cultivated through culture.

Crucial to mystery fiction is Aphrodite's adulterous alliance with Ares, god of war. As Paris explains, today neither Aphrodite nor Ares are prime movers of war to the detriment of the psychic health of all of society.[31] Today Apollonian dominance in technology, distance, and disembodied order has triumphed over the somatic fierceness of Ares. What has largely replaced face-to-face combat is war fought at a distance through unmanned machines controlled from air-conditioned bunkers thousands of miles away.

Perhaps as a result, Aphrodite's modern role in warfare is almost wholly negative and destructive. Rather than the sacred prostitutes of Aphrodite who might have gently restored wounded bodies and psyches, Paris points out that rape is used as a mode of terrorizing and crushing a population in defeat. Such a situation brings into question the complex weave between mystery fiction and war. After all, mysteries almost always center on murder, while in wars the same nation that punishes murder sends soldiers out to kill, and thereby encounter terrifying gods and goddesses.

Here I want to suggest that mystery fiction, born in its modern form shortly before World War One, is a trickily fictional attempt to link our ancient history of hunting and being hunted with our modern condition of war. Strategies of the hunter are developed even in modern technological warfare by the snipers and those who direct drones. Yet modern war is distinguished by its capacity to annihilate populations, not just those who choose to fight. It is therefore particularly traumatizing to human culture in threatening its very survival. When survival is at stake, the trickster surfaces.

In her book, *Twentieth Century Detective Fiction* (2001), Gill Plain marvelously depicts how far Agatha Christie's relatively clean, sanitized, *and recoverable* dead bodies seem to be compensation for the horrific mutilations of corpses in World War One's trenches.[32] Was her work in particular, and the "cozy" genre she helped to mold, attempting to ameliorate the terrible absence of those millions of bodies interred in mud?

I would like to propose that the whole mystery genre may be attempting to compensate for, or rebalance, the cultural psyche in a time when war seems to efface the meaning of an individual death. For above all, mysteries make death meaningful, and therefore solvable at the level of self-conscious fiction, as I argued in Chapter 1. Mysteries convert death, in all its many occasions—from old age, illness, accident, nameless acts of war, to conscious and preventable murder. They therefore return the deliberate killing of a human being to the individual and embodied plane of Ares, lover of Aphrodite.

As a metonym is language that works by being part of what it signifies, one could say that at some deep level mysteries are metonymic substitutions for war, for physically taking up arms against an enemy. Specifically, Aphrodite and Ares are deceiving her husband, the iron-working god, Hephaestus, whose forge is arguably the mythical source of the drive to mechanize war with ever more sophisticated weapons of mass destruction.[33]

Tricking Hephaestus, then, is an erotic, embodied, and personal gesture against convention (marriage), one that arguably undermines him as the propensity to make machines (of war). When unproblematically allied, Hephaestus and Apollo will sponsor the machines and disembodied reason that pervade modernity. These two principles have come to dominate warfare and to degrade Aphrodite's divine powers into yet another dehumanizing weapon, rape. In a society that has historically repressed the capacity of sexuality to be both feminine and civilizing, weapons, ingenuity, and warrior courage all seem to have *lost touch*. They have cast aside the playful connection to Aphrodite that might mitigate their dark potential. How might a case study in women's mysteries explore this sad condition?

Aphrodite returns as also a trickster. She deceives and causes pain, yet her erotic bounty remains sacramental in uniting bodily desires with the survival of the soul.[34] Whether it is in the capacity to heal the warrior to come *home* from battle, or to offer the fictional detective that instinctual *knowing* that similarly heals the realm of the mystery from continued killing, it is time for a case history to explore her invigorating divine powers.

Case History (5): *Trophies and Dead Things* (1990)
by Marcia Muller

Sharon McCone, investigator for All Souls Legal Cooperative, San Francisco, is feeling the absence of Aphrodite. Having fallen in love with the client of a previous case, her beloved has left her in order to tend a sick and estranged wife. He may never return. Fearing empty hours, she is keen for a new assignment. Her task is to find the unknown heirs of a man recently killed in one of several apparently random snipings.[35]

Sharon's friend, Hank Altman, who is also her boss, is the deceased Perry Hilderly's lawyer, perhaps because they had known each other in Vietnam. Yet, shortly before he died Perry changed his will without consulting Hank, who, learning of this, says, "I don't know *what* I know anymore."[36] Despite appearances, Hank's comment is prescient, for he does not know that he *does* know the link between the sniper's victims. He also almost becomes a victim before Sharon figures the case out.

Hilderly's four mysterious heirs share a link to the anti-Vietnam War protests and Free Speech Movement of the 1960s that were centered in San Francisco and Berkeley. The youngest heir is the daughter of a protester who killed herself, and the other three are deeply marked survivors of an event so traumatic that they refuse to disclose it. One of these is a particularly dislikeable lawyer who makes gruesome fetishes out of "dead things:" feathers, bones, and metal. Sharon recalls a verse from John Webster, a dramatist of the seventeenth century specializing in murder, revenge, and extreme emotions: "vain the ambition of kings/who seek by trophies and dead things/to leave a living name behind/and weave but nets to catch the wind."[37]

Trophies and dead things appear to stem from the loss of Aphrodite to more than just the investigator. War, in which Ares gave way to Apollo in Vietnam, was without a bond to Aphrodite's civilizing arts of love for so many veterans. It is this dehumanizing war that is behind the snipings. Far from being random, they are the crazed revenge of someone who cannot escape his experiences.

Moreover, Hilderley's three heirs from the 1960s have been wrecked by their inability to keep their ideals of a more just society in touch with Aphrodite's celebration of the life of loving bodies. Two of them,

Libby and D.A., were imprisoned for attempting to bomb a military base. The third, the maker of trophies from the dead things, was the planted government agent who betrayed them. The fourth heir proves to be his daughter, whose mother he pushed into committing suicide lest his indiscretion impede his "vain ambition."

Aphrodite is multiply betrayed in the divorce of Ares from his goddess lover as the conspirators lose touch with the value of an embodied life. She is traduced by the spy who destroys the joy of his relationship with a woman, whom he pushes into despair. And, of course, Aphrodite is betrayed by a society that uses control and violence at the expense of attempts to connect erotically. Fortunately, Sharon's loss of the goddess is not her rejection of divine feminine sexuality. Her previous relationship with police lieutenant Greg Marcus ends with both of them respecting the *knowing* of the other, and so they are able to make an alliance in detecting.

Also Sharon instinctively pays tribute to the polytheistic psyche. She says early on that she needs to know the truth, a statement that resonates with Jess, daughter of the suicide victim of the 1960s conspirators. However, the truth is not the same as the mere facts of the attempted bombing, nor is it that all the recent sniper victims knew each other in a bar in Vietnam twenty years previously. This "truth" requires the presences of several goddesses to be acknowledged. Sharon's consciousness of the depressive feelings caused by the departure of Aphrodite in her own life enables her to *know* what the harsher rending apart of the goddess did to Jess's mother. Yet she also embraces the Athena mode of close working with men, which again gives her insight into the horror of Hank's own experience in Vietnam, enough to sense the desperate absence of Aphrodite in some veterans.

In addition, Sharon has another goddess within: Artemis, whose hunting prowess and drive for purity of life she embodies in insisting on fitting shards of the past to the fractures of the present. Only then can she deduce that the most wounded survivor is the most deadly, getting his revenge without hope of a goddess. Unlike this killer, Sharon knows both Artemis and Hestia, the hearth maker, gaining insight at the hearth she shares with her friends. For now it is primitive and sometimes death-bringing Artemis to whom we must turn.

Hunting Artemis

Divine hunter, Artemis sanctifies solitude and living in close
relation to non-human nature.[38] She therefore grants access to a wild
and primitive feminine that is entirely independent of any other
goddess or god. She is the feminine radically autonomous and self-
sufficient, one almost lost sight of in Sky Father dualism that posits
gender as dialogical. By contrast, Artemis is an Earth Mother who
connects all nature, siting humanity as creatures with a natural habitat
of the wilderness.

Unsurprisingly, Artemis has been little heeded in a patriarchal
society in which extinction of species is regarded as of little importance.
Not so to Artemis whose hunting nature is one woven into the
sustenance of wild animals. Indeed Artemis is profoundly implicated
in mysteries of life and death. By forbidding a hunter to wound an
animal rather than kill it and end its suffering, Artemis brings death if
it is necessary to sustain the primal purity of life she stands for.[39] It
follows that although a virgin and never a mother, Artemis accompanies
women in childbirth when death and new life are intimate with each
other. With courageous insight, Ginette Paris suggests that Artemis
offers abortion as a sacrament, a necessary infliction of death when the
alternative is to maim the living.[40]

Artemis is fierce in the protection of instinctual life, a protection
that includes the mysteries of death. She therefore drives martyrs,
suicidal heroism, and sacrifice. This extends even to human sacrifice
in the deep archaic past as is suggested by the myths surrounding
Iphigenia. In her myth, Iphigenia is a Greek princess ritually killed
by her father, Agamemnon, to enable him to proceed with the war
against Troy.

Yet Iphigenia is not simply a tragic and passive victim of
patriarchal aggression. Paris shows that Iphigenia is associated with
Artemis herself, not just as one sacrificed to her but also as containing
some of her divine, primal energy.[41] Hence it is not so surprising that
unmarried and childfree detective Victoria *Iphigenia* Warshawski,
created by Sara Paretsky to haunt the Chicago mean streets, is
indefatigable in defense of the weak, and bears the scars of her own
sacrifices on her body and psyche.

Above all, Artemis is a goddess of knowing the feminine as divine and as independent of any other reality. As Paris puts it, Artemis enables a connection to a wild feminine beyond possibility of domestication and of being trapped in social conventions. Every fictional detective who steps outside those compromises demanded by an imperfect system of justice is an Artemis picking up her bow and making for the mountains. The sleuth who needs to know in order to fit her nature to the nature she is connected to outside of conventional bounds is Artemis who "know[s] the art of preserving within [herself] a force that is intact, inviolable, and radically feminine."[42]

So Artemis in detective fiction is one whose drive for purity of life is uncomfortable for those around her, yet this drive forces her to penetrate the mysteries of life and death. Such an uncompromising goddess at work incarnates the need to know the truth about the murder. It means knowing not just who did it or who confessed to it, but who and what set of circumstances was ultimately responsible. Like the Artemis who will bring death if that is what it takes to preserve wild primitive living, such an attitude on the part of the sleuth can be very dangerous to those around her.

A detective who must continue the hunt for the truth to the point where it satisfies a divine appetite for psychic justice is one who may deliver death as part of her pursuit. In *The Sugar House* (2001) by Laura Lippman, Tess Monaghan appears an unlikely Artemis, until her father confronts her over the smoking ruins of the family home and blames her persistence for the burning house and the dead body inside.[43] Perhaps even more stark is the example of Kinsey Milhone in *K is for Killer*.[44] At the end of her hunt she makes a phone call that she knows will lead to another murder.

Case History (6): *K is for Killer* (1994) by Sue Grafton

> The victims of unsolved homicides I think of as the unruly dead: persons who reside in a limbo of their own, some state between life and death, restless, dissatisfied, longing for release.[45]

Kinsey Milhone, who is deeply attached to her home town of Santa Theresa in California and who lives in a tiny apartment, would appear to be an unlikely Artemis. However, her opening concerns for victims of unsolved murders as the restless dead reminds us of Artemis as protector of the boundaries between living and dying.[46] Milhone works and lives alone, although she has a strong connection with her landlord, retired baker Henry Pitts. Solitude is not only necessary to her personally but is also vital to her detecting. A young woman with a P.I. License, she can and does ask questions not available to the institution of the police. Moreover, sitting alone with the facts of the case spread out on index cards is a vital process in her pursuit of the guilty.

Lorna Kepler died alone, and when finally found her body was too decayed to ascertain cause of death. After a pornographic film, starring Lorna, is sent anonymously to the family, her mother asks Kinsey to investigate. Kinsey soon discovers that Lorna had an ambiguous relation to the darker side of Aphrodite. Working as a high class, highly paid escort, Lorna was unusual for a woman sex worker. She stayed in control of her professional activities and was making a lot of money. She was, in this sense, *virgin* Aphrodite: self-contained, and in some sense psychologically inviolate.

Yet Lorna was also Artemis, living alone by choice in a cabin surrounded by trees, forging a mysterious bond with a friend's huge dog, and mentoring a younger, more naïve prostitute, Danielle Rivers. Kinsey/Artemis is looking for clues to Lorna's death by following her on the mysterious paths from the disconnected parts of her life: from an apparently loving family to a mundane job in high class bars and hotels to the friends who found her hypnotic, special, and yet were told little of her life.

Through her investigation, Kinsey and the reader get a sense of Lorna as a complex personality seeking a purity of life, and with a gift for friendship. Beautiful, she incurs the jealousy of her landlord's pregnant young wife, who bugs her apartment. What the eavesdropper hears is Lorna telling her self-absorbed spouse to be understanding of a woman's burden with childbirth and caring for an infant. She tells him to help more in the house. Here is the Artemis woman who protects women around childbirth. Perhaps it is significant that Lorna is killed just when she was about to marry a man whom Kinsey is sure is highly placed in organized crime.

By dying before her marriage, Lorna distances herself from possible identification with Persephone, the daughter of Demeter who was abducted by Hades, god of the Underworld. Demeter obtains her daughter's release, but not before Persephone has ingested enough of the Underworld to cause her to return there for part of each year as Hades's queen. Although Lorna certainly tastes the Underworld's fruit in her thriving one woman escort service, she was not abducted or forced down into that world. She chose to go there and seems to have chosen to remain in marrying into the lucrative and highly dangerous orbit of the mafia.

Such a position as Queen of the Underworld was not to be hers. Lorna has taken care not to let her various domains overlap. Yet a collision occurs when her boss from her day job becomes a client, and she and Danielle witness how corrupt he is. Lorna is killed; later Danielle is horribly beaten and dies of her injuries. Two moments link Kinsey to both murdered women in an intimacy that is of Aphrodite and Artemis. Visiting Lorna's night owl friend, Hector, and his dog Beauty who mourns Lorna, Kinsey seems to provoke an otherworldly despair in the animal.

> The howling became a low cry, filled with such misery that it broke my heart…
>
> "I feel bad," I said. "I was wearing those jeans when I went through Lorna's files."[47]

Kinsey has taken on Lorna's smell and this makes the bereaved dog experience hope followed by renewed despair. Beauty becomes a real character in this novel. Her relationship to Lorna and her depression at her vanishing is a vivid contribution to Kinsey's archetypal goddess-activated detection. Not only do Lorna and Kinsey embody Artemis in their primal relationship to Beauty, the dog's name also reminds us of the beauty of Artemis as primal feminine, virginal nature. Crucially, this moment is one in which Kinsey, huntress for a truth that will restore a purity to her client's (Lorna's mother) life, and increasingly to the memory of Lorna, takes on Lorna's scent, in order to understand her in the sense of standing on her ground of being. Indeed, by literally inhabiting Lorna's body, Kinsey brings something of Lorna to rebirth.

Kinsey also experiences the goddess through Danielle, who shares Lorna's essentially solitary pursuit of financial stability by unconventional means that thrive on primal instincts. Danielle turned to prostitution because it pays more than hairdressing. She insists on giving Kinsey an expert and flattering trim. In almost all the novels Kinsey remarks that she only ever cuts her hair with nail scissors. True to an Artemis trait of a feminine as natural and autonomous, Kinsey normally dresses for no one but herself.

So, to be in the hands of Danielle cutting her hair into a delicate and feminine shape is an extremely rare moment of Kinsey experiencing feminine Eros and expression of sexual beauty. Danielle momentarily awakens Aphrodite in her, which is perhaps why Danielle's death is inconceivable to Kinsey. Literally, Kinsey cannot conceive it, she cannot let it come into her body by accepting the news. It is in this crisis of knowing and being that Kinsey phones the mafia don, claiming to represent Lorna's fiancé, and gives him the name of the killer of both women. What Kinsey knows through her Artemis hunt is that this man is guilty, yet cannot be prosecuted for lack of evidence. Refusing to accept the death of Danielle who had become a friend, Kinsey is flooded by Artemis as avenger of vulnerable women in order to restore purity of life.

Almost immediately Kinsey is appalled at her action, and tries to stop the revenge of the mafia underworld for the stealing (by death) of their queen. She is too late, mainly because the guilty man prefers attacking her to listening to her. Finally, Kinsey is left to ponder the eruption of primitive violence in her own being.

> As for me the question I am left with is simple and haunting:
> Having strayed into the shadows, can I find my way back?[48]

It is the nature of Artemis to stray, to venture into a wilderness where there are no conventional paths, outside the law. However, Kinsey knows that Artemis is an archetypal need in her that cannot go unrestrained in its fiercely independent and sometimes death-dealing feminine nature.

The Mysteries of Hestia's Hearth

Hestia is the goddess of the center; center of the self, of the home and of the Earth.[49] She is the hearth fire that makes a home. "Hestia" means the goddess and also the flames of the sustaining hearth at the center of the earth.[50] Therefore Hestia is where the family makes its home. She also is a personal sense of center: the ability to be at home in our own bodies and souls.

Such a powerful *necessary* goddess of well-being lends herself to stillness, to creating the home that nurtures. So in what sense can Hestia be found in mysteries with their energetic quests for difficult truths? One answer lies in Hestia as protector of the stranger, for the hearth is sacred. Once a stranger is accepted at the hearth, he or she is under the goddess's protection.

> …Hestia protects, receives and reassures. When a stranger was invited into the area of the hearth, he was protected, for this place was sacred.[51]

Hestia is also patron of mysteries that center on the home and family, which is either itself threatened by the crime; or where someone, including strangers, accepted at the hearth/heart of the home is in danger. "Cozies" typically have a female sleuth who, although not professionally involved in law enforcement, is drawn into the quest because of concerns for family or the stranger at the hearth. Here we have figures such as Lucy Stone, wife, mother, reporter, and sleuth, from the pen of Leslie Meier, the bed and breakfast mysteries of Mary Daheim, and caring Annie Darling, bookstore owner whose passion for justice in her immediate family and community is drawn from the imagination of Carolyn Hart.[52]

Hestia sleuths defend the family and the stranger at their hearth as the centering energy of the home. They *cannot* stand by when violence or crime threaten to destroy it. A particular subset of the cozy is the food mystery, in which the detective is primarily a cook and recipes appear in the book along with the quest story. As Ginette Paris astutely notes, food is the *matter* of divine Hestia when it serves to unite the family.[53] Hestia is the communal, community-making aspect of food, while Demeter stands for the harvest.

So, unsurprisingly food mysteries frequently involve a cook with a catering business who stumbles across dead bodies when her hearth, or catering event, is meant to enact Hestia for family and close knit community. Diane Mott Davidson's detective, Goldy Schultz, has a cooking enterprise run from her own kitchen.[54]

Katherine Hall Page's Faith Fairchild lives in a small New England town where she combines the role of mother, wife to a clergyman, and caterer, while Isis Crawford writes about catering sisters, Libby and Bernie, who live close to New York.[55]

Joanne Fluke's very popular Minnesota series, with cookie store owner Hannah Swensen, perfectly illustrates how the Hestia sleuth is not confined to being a traditional homemaker.[56] A romantically minded spinster living alone, Hannah is nevertheless intimately bound up with family and community through the act of solving mysteries. Additionally, she is family-centered because her cooking is an enactment of familial connection extended into the community of significantly named Lake Eden. Moreover, her cookies and cakes become a material medium for drawing her into a murder, enacting the desires of those around her for sweetness. Sleuthing, for Hannah, restores Lake Eden to a mythological Garden of Eden by centering and discovering what is necessary for home to be reconstituted. Perhaps this accounts for the large dose of sugar in the portrayal of tensions within a small town named for paradise.

The powerful presence of Hestia in cozies does not preclude her divine energy from extending to other mysteries by women writers. Centering a family, or a group making a new kind of family, is arguably innate to women's detective fiction. Even loner Kinsey Milhone has a familial relationship with Henry, her aged, yet still handsome, landlord. V.I. Warshawski and Sharon McCone each find that many of their professional cases invoke Hestia for their clients, along with drawing familial dimensions into their own lives. So in *Trophies and Dead Things* Sharon takes on a project that appears distantly linked to her friend, only to discover that the people she is most connected to, her new family, are in danger.

Sara Pareksy's V.I. appears less able to enjoy the comforts of Hestia than Sharon, who eventually marries her long-time partner.[57] V.I.'s parents are dead, and she is estranged from living relatives. However, memories of parental love remain a vital centering force to her. They make sense of V.I.'s drive to solve crimes that have both a political and a familial aspect, because solving the crime makes the hearth fires of home again ready for kindling.

Indeed, these professional female Private Investigators share something of Hestia's sacredness in their contract with clients. In these mysteries the client is a stranger taken to their hearth. The client is sanctified and defended, sometimes even after death. For example, in *M is for Malice* (1996) Kinsey is determined to defend the honor of the long lost brother she was hired to find by his uncaring, Hestia-defying family.[58] She finds him, but when he is swiftly murdered, she refuses to be stopped by what has become a violation of a sacred trust.

Hence, Hestia detectives are not confined to home-based amateurs, nor do they always work within a group. Even the lone sleuth will often invoke Hestia by herself being the center of a web of friends. She then makes the *solution* a re-igniting of the home building hearth fire. As Paris explains, Hestia is opposite to Hermes in that she centers the home while he rushes about in a flurry of communicative travel.[59] In fact, both goddess and god are important to the modern detective. While Hermes may be found in the swiftness of electronic media, Hestia is a knowing based on centering, on the fires that make a group into a family that do not have to be (indeed often are not) the traditional nuclear family. Knowing Hestia is to feel at home in one's body, psyche, (modern) family, and community. It is to *make* one-self a home in detecting the ultimate threat to the hearth as sacred: the killer.

Therefore above all, Hestia inhabits women's mysteries in the aspect of knowing that finds the center in the solution to the murder. Finding the solution centers and re-ignites the hearth; the community is reborn and the wasteland healed. Hestia here embodies, and is also the object of, the quest. The detective has to *know* because only solving this crime will re-start the hearth fire. The stranger lies dead, and so the sacred fires need re-kindling in a heroic quest for the center, the home, of this person, or this group and this Earth-centered nature.

Case History (7): *A Catered Thanksgiving* (2010) by Isis Crawford[60]

> It was two days before Thanksgiving and five members
> of the Field family were huddled around the fireplace
> in the study off of the living room. It was a dismal
> space... A flickering overhead light did little to dispel
> the gloom of the late November afternoon. Each one
> in the room was wearing his or her coat.[61]

Here at the start of the novel Hestia is vitally absent. Her absence is *vital* because this family has no vitality without its hearth fire; this family has no home. This fireplace is empty; the room dark and cold. The Field family patriarch, Monty, has hired Libby and Bernie Simmons of the shop, A Little Taste of Heaven, to provide some divine intervention by catering their Thanksgiving dinner. Unfortunately Libby and Bernie are not feeling very thankful either. Their disabled father has taken himself off to relatives in Florida and their family hearth risks being untended.

Even more disastrous is the blizzard that begins as they drive to the remote Field house. It is indicatively situated next to a concrete bunker where the family business of fireworks manufacturing is still in operation. Fireworks, rather than hearth fires, represents the Field family under the miserly rule of Monty. They provide explosions without any warmth, any trace of making home. However no one expects the selfish old man to die of an exploding turkey on Thanksgiving Day. Horrified, Libby and Bernie discover they are trapped in the icy house with the feuding family and a dead body. Until the phones begin to work and they can call their ex-cop father, they have only each other and the optimistic possibility that only one member of the family is a killer.

Family business is the key to this mystery in more ways than one. The Fields have made a lot of money with their fireworks. Monty kept most of it for himself, earning the resentment of his brothers and adult children, not to mention a new and expensive wife. Moreover, in that family business brings Libby and Bernie to the house with the fatal turkey, means that these very unlike sisters cannot separate work and home life. Furthermore, in this case family business also extends to the police. Not only was their father, Sean Simmons, injured in the line of duty, he was also forced out of the police department for political reasons. One factor making him vulnerable to unfair treatment was that he once attacked Monty Field for cheating his wife (now dead) over a catering event for the business now run by his daughters.

Food prepared by Libby for A Little Taste of Heaven is delicious and innovative. It draws family and friends to their hearth as well as customers who may or may not have hidden agendas. Through catering, Libby and Bernie are the center of a network that extends via ex-cop Sean into his friends still in the police, the boyfriends of both women, and those people in this tight-knit community bordering the city who want to use their distinctive food to enhance social status.

Trapped in the cold house with family members continuing to be murdered, Bernie and Libby are forced to explore familial space; their own and that of the Fields. Familial territory proves to have hidden perils in relationships both familiar and secret, and also in the unexplored material divisions of the house itself. While searching for missing Geoff Field, the sisters discover a concealed staircase leading to bedrooms once used by a servant and her son.

> She tightened her grip on her knife, just in case Geoff was up there waiting for her but she didn't think he was. The space felt empty, devoid of life. She wouldn't be able to explain to anyone why she felt that way, but she did, and by this time she'd learned to trust her instincts. It was when she didn't that things usually went wrong.[62]

Bernie insists on following her embodied intuition into this dark space that is "devoid of life," thus most lacking in Hestia divine energy of the home-hearth. Libby protests but will not let her sister face danger alone. Together they discover the violation of Hestia by Alma, who has herself found no secure home at the Field's when she was seduced by the patriarch, Monty, and gave birth to a son. Denying paternity, Monty had illegal immigrant Alma deported, so betraying both familial responsibilities. Bernie and Libby share enough Hestian fire from their family and food preparation to sniff a motive for murder in such behavior in these cold apartments.

Roberto, the disowned son, is the obvious suspect for who has been murdering the Fields. However, when Libby and Bernie discover that the whole house has been rigged to explode, it is the legitimate daughter, Melissa, who proclaims that she is "El Huron," the vengeful killer from the dark. In a dramatic twist, the sisters learn that she, not Monty, was responsible for deporting Alma and making it look as if Roberto was Monty's son. As she is *apparently* suffering from multiple personality disorder, Clyde, Sean's cop friend, says that it is also possible she is pretending, and did it all for the money.

What is certain at the end of *A Catered Thanksgiving* is that the Fields were a fractured family. Cracks in the psyches of its members extend to the possible detonation of their house. Devoid of Hestia, relationships turn sour or lose so much meaning that fantasy ones take over. Libby and Bernie cannot help the Fields because Monty's dominance literally kills him when he insists on a ritual inspection of the cooking turkey. Unlike the victim, food for the Simmons' materializes and transmits the loving Hestia hearth. Reflecting upon how much he enjoys his daughters' cooking, Sean Simmons compares himself with Alma, who has lost her son to Melissa's murderous rage. He knows himself to be lucky, despite his unfair professional treatment.

Libby ends the novel by summoning family, friends, and boyfriends to a late Thanksgiving Dinner, "because... we have a lot to be thankful for."[63] Hestia enables the quest and the knowing of family that proves crucial to solving this mystery. Where Libby, as the cook who connects through her imaginative ways with food, is Hestian, Bernie, with her taste for fashion, offers something of Aphrodite to her sister. Although often resisted, Bernie, through sharing with Libby the preparing and serving food, takes on enough Hestian fire to share her sister's intuitions about dysfunctional families. After all, it is Bernie's taste for Aphroditean adventure that gets her home-loving sister on the icy roads to the Field house.

Aphrodite leads Hestia outside to try to start hearth fires through their catering. Together the sleuthing sisters provide a Hestia knowing of the familial center that proves essential to stopping a mad daughter from continuing to kill.

It is time to consider the detective who wants to work within the system for the good of the whole community. For Athena too, inhabits this trickster genre, in the woman who likes to work within male structures.

Athena's Troubled Cities

> Athena was the beautiful warrior goddess who protected her Greek heroes in battle. She was the goddess of wisdom and crafts, a master strategist, diplomat, and weaver, and patroness of cities and civilizations... She sided with the patriarchy in casting the deciding vote to free Orestes, who had killed his mother.[64]

Also pointing out that divine Athena never acknowledges her mother, Metis (preferring to consider herself as wholly Zeus originated),

Maureen Murdock's portrayal of Athena suggests a feminine wholly imbued in the patriarchal system. On the other hand, Christine Downing, in *Women's Mysteries*, notes that Athena is particularly interested in the fate of wily Odysseus, the most tricky of all the Greek heroes at Troy.[65] In Homer's *Odyssey*, Odysseus is also the one given to extraordinary adventures on his long way home.

Indeed, Downing says that Athena calls on women to take on the male world in male terms, seeing everything in terms of the interests of the community.[66] Athena is here the embodiment of arts and insight: "the goddess of clear vision and artistic power."[67] It is therefore hardly surprising that Athena is particularly well represented amongst detectives created by women writers in Linda Fairstein's sex crimes prosecutor, Alexandra Cooper.

> Mike and Mercer came from backgrounds as different from mine as one could imagine, but we had the same respect for the criminal justice system and the same value for the dignity of human life. Both of them had helped train me – every bit as much as the lawyers from whom I'd learned – in the art of investigating cases, in the search for the truth that characterized the way a great prosecutor's office worked.[68]

As a lawyer firmly dedicated to making the masculine-dominated justice system work, Alex Cooper is Athena, whose closest comrades are male cops. Where Alex differs from Athena is in the goddess siding with the patriarchy, for example in deciding *for* Orestes after he killed his mother, Clytemnestra, and her lover (who had previously murdered his father). Clytemnestra initiated the cycle of revenge for the sacrificial killing her daughter, Iphigenia, by her husband, Agamemnon.

Alex does not side with patriarchy in her work prosecuting the (mostly) male criminals who have raped or sexually assaulted (mostly) women and children. Indeed, *Night Watch* (2012) opens with Alex risking her romantic relationship with a Frenchman, Luc, when he takes the side of a socially important man accused of rape. On the other hand, as James Hillman points out, the important role of Athena is that her finding for Orestes makes a place for the Furies *within* the divine order.[69]

Athena here really is a lawyer in the modern sense, for her persuasive speech incorporates revenge into a system of order by which the community can live and flourish. As such, she allows Alex to uphold the father's law that fortunately has been changed to take account of the true horrors of sexual crime. In this sense Alex is patriarchal Athena whose defense of the father's law is essential to her role of making a place where the potentially unending furies of revenge can be assuaged within the city or community.

Alex Cooper precisely embodies such an exponent of skillfully woven words dedicated to stitching together the body politic of her city. Here also is her characteristic love of New York which is enacted in plots that explore and bring back to her ordering communal influence the city's deep fabric such as the water pipes beneath the streets in *Bad Blood* (2007) or the literary legacy of Edgar Allen Poe in *Entombed* (2004).[70] In her devotion to the history of New York she reminds us of Athena's protection of Athens.

Athena remains virgin despite her intimacy with masculine order. Similarly, Alex, while erotically attracted to blue collar detective, Mike Chapman, places her professional relationship with him above its romantic potential. In fact until the most recent books, she, most Athena-like, sees a love affair with Mike as threatening her power to work effectively as Athena/lawyer.

> Somewhere along the line, Mike Chapman had become my closest friend…If I gave any thought to dating Mike, I knew that Battaglia would relieve me of my position. He wouldn't allow the impression that a top detective was closing cases or eliciting confessions because he was sleeping with a supervising prosecutor.[71]

Alex habitually chooses Athena over Aphrodite; to be inside the patriarchal system where communal values of political appearances threaten to prevail over the personal integrity that marks her relationships with cops, Mike and Mercer. The reward for such sacrifice is for Alex to enact what Hillman describes as Athena's "institutional mothering."[72] Her nurturing and defense of vulnerable victims of sex crimes embodies the sacred mothering of the state or of the justice system. It is an Athena aspect bound up with her role of enacting the divine necessity of reason in Alex's skills and strategies as a prosecutor.[73]

Other female authored sleuths contain Athena characteristics. Even loner, Kinsey Milhone, has an Athena-like willingness to work with the police, sometimes. Another example occurs in *A Lesson in*

Secrets (2011), where Jacqueline Winspear's psychologist-detective, Maisie Dobbs, has an Athena streak when she agrees to become a British intelligence agent in order to combat a Nazi spy ring in 1930s England.[74] Mostly, however, the fictional sleuth prefers to work alone, with a well-founded distrust of pervasively masculine institutions like the police.

It is not so much that the police in these mysteries prove venal, but rather that they lack, disrespect, and in some cases feel threatened by, feminine divine powers of detecting. Their priorities and parameters are too limited to pursue the *wholeness* of being and knowing demanded by the genre. Such a deficiency in institutional order is still true even when the fictional sleuth is actually a member of the police force, as Kinsey McCone recalls of her time as a cop, in Sue Grafton's *O is for Outlaw* (2001).[75] Hence the ubiquity of the maverick cop as a staple of the genre. She, or he, is always in danger of being ejected from the masculine pantheon of law enforcement. The maverick or genius cop, frequently exceeding police rules, is more often the creation of male writers, such as Ed McBain, Elmore Leonard, and the more genteel, clue-puzzle variety, such as Colin Dexter's UK, Oxford based, Inspector Morse novels.[76]

As a generalization, women writers have tended to prefer sleuths working external to the law, whether licensed Private Investigators or amateurs in cozies. Lawyer-sleuths are an exception. For the pursuit of justice in a courtroom, Athena's strategies and weaving ability with words can be a divine vehicle of healing, reconciling individual hurts with Athena's communal values. On the other hand, as Alex Cooper discovers, Athena's divine reason and skillfulness within the masculine system is no protection against surprises.

> So even in the realms where Athena and rational strategy are most called upon... there are obvious and tragic strategic failures because too many other dominants have been left out of the calculations.[77]

These other dominants may include less rational qualities of ambition, greed, and desire. A good example of the challenge to Athena comes to Alex Cooper in *The Deadhouse* (2001), a novel that begins when her good sense over a trick to catch a violent man is ignored.[78] So Athena here cannot prevent an unlooked for murder.

Case History (8): *The Deadhouse* (2001) by Linda Fairstein

It was hard not to smile as I watched Lola Dakota die.[79]

Despite Athena's attraction to stratagems to defeat an enemy, here a wife abuser, Alex Cooper nevertheless decided that this particular sting to trap a potential killer was too risky. Her warnings pushed aside, Alex is proved correct when shortly after the staged murder of academic, Lola Dakota, she is found genuinely dead in the elevator shaft of her New York apartment building. Alex teams up again with homicide cop Mike Chapman and Special Victims detective Mercer, in order to pursue the criminals. Her role is to prepare the case for trial in court.

Opening *The Deadhouse* with the trap for Lola's violent husband recalls the trickiness of the mystery genre itself. What on the level of narrative realism is a plausible strategy with a wily wife beater, is on another level is a display of the self-conscious reflexivity of detective fiction. The genre is characterized by its inbuilt sense of a sophisticated readership with experience of the ritual repetitive aspect of mysteries. Indirectly referring to its status as fiction, *The Deadhouse* enacts this further evidence of Athena's love of contrivances expressed within the system.

After all, the mystery genre itself is, in its knowable and predictable codes, part of the structures of society and psyche. Athena finds a home in this particular city of the mystery in which regular patterns can be discerned if manifested in individual expressions. Arguably, Athena embodies here what distinguishes almost all fictional detectives: that primal sense of quest for a justice that can restore communal values. James Hillman notes that Athena has an association with the goddess Persephone.[80]

Abducted by Hades from her mother in Demeter's fertile valleys, Persephone is only restored to the upper world after Demeter's grief causes a terrible famine. Yet, having eaten some pomegranate seeds while in the Underworld's darkness, Persephone must return there for several months of the year. During her daughter's absence, Demeter makes winter on the Earth. Perhaps Athena is allied to Persephone in her accommodation with the male order through becoming Hades' Queen for only a few months each year?

Two aspects link Alex Cooper with Athena in association with Persephone. Given the possibility that Persephone's abduction refers to rape, Alex-Athena is a compassionate yet pragmatic supporter of justice and restitution for such victims. She is not their avenger as Artemis might be. Alex's progress in all the novels is to distill rational discourse from a traumatic event. For example, she guards the male order from false accusations by women as well as transmutes the Furies of violation into their place, making a *divine* order of the justice system.

Secondly, like Persephone, and even more like Demeter, in aiming to rescue Persephone, Alex herself frequently descends into the underworld, usually to try to bring back a female victim. Here "the deadhouse" is an abandoned morgue on an island off New York once used for those dying of infectious and incurable diseases.

> "The River Styx, Lola used to say this was. Souls crossing over from the realm of the living on their way to hell. To what she called the deadhouse."[81]

Alex first visits this gruesome relic safely in the company of Mercer and Mike. However, later she finds herself forcibly returned to this place of death after being abducted, like Persephone by Hades. She is snatched by the murderer. Now trapped in a hellish underworld of masculine unreason, she has to plead for her life far from what tries to be the rational order of the courtroom where she usually speaks her Athena reason. Yet it is her Athena affinity for strategy and devices that ultimately saves her.

The murderer wants a treasure map she is concealing on her person. Unable to overpower her male attacker, she needs to summon help by stealth. So she persuades the murderer to speak to Mike Chapman, in a message that carries a double meaning. Again, her device appears to fail; but, fortunately, the message reaches its destination and Alex is rescued. In the meantime, notably, she has already saved herself by running away into the rocks. Wildly pursuing her, the murderer is swept away in this particular River Styx.

Here Alex's contrivances through speech prove effective. Like Persephone she makes a deal with the darkness in order to leave it. Also like Persephone the taste of the underworld never entirely leaves her. Earlier in the book, Alex expresses distress that a murder victim's body has been left exposed to the elements. Unlike Artemis of the wild woods, Athena and Alex need the bodies to be brought into the city for proper burial. Imprisoned in the deadhouse, or ex-morgue, Alex discovers the body she has been seeking. Her rescue will enable the murdered young woman to be taken home.

Goddess myths inhabit mysteries by women in ways that offer greater imaginative possibilities to gender and genre. It is time to get closer to these divine beings, to track their pathways to the sacred through the many exciting possibilities offered by the mystery novels of women writers.

CHAPTER THREE

HESTIA
DETECTING HEARTH AND HOME

Introduction

Hestia, goddess of hearth and home, might appear an unlikely divinity for the mystery's quest for truth. In the Greek pantheon, Hestia is paired with the enquiring, mercurial Hermes, in the sense that Hestia inhabits the hearth, stays at home, while Hermes is darting about delivering messages.[1] It is he who uses language trickily in ways familiar to criminals and detectives. On the other hand, as James Hillman says, at the end of each day of legal cases the judge would leave a written account of the proceedings on the altar of Hestia.[2] If Hermes is the unreliable patron of trickiness in words and father of hermeneutics, Hestia conversely guards contracts. Hestia's relationship to the word is one of fidelity to meaning that makes writing a core ingredient of a centered sense of home, in the city, amongst family, and for the self.[3]

Hence, Hestia is found in the sacred commitment between sleuth and those for whom the investigation is undertaken. Such a bond occurs whether the witnessing of truth takes the literal form of a written contract or not. So, "centering" and stabilizing being via the quest for the *right story* is a clue to Hestia's role in detecting to re-make home. In some of the mysteries discussed below the amateur sleuth is forced to solve a situation that is blighting her own hearth. Hestia is here the object of the detective's quest, while providing a crucial dimension to knowing. By finding out what really happened, hearth fires are both (re)discovered and re-ignited.

However, Hestia also resides in detective fiction by women in negative forms. Some detectives are spurred on by the mystery of Hestia's absence in a family or in a workplace that is cold and uncentered. Detecting the ways in which Hestia is absent draws upon a mode of knowing based on familial warmth and trust, in order to solve by dissolving the uncanny disappearance of these qualities. Yet also, Hestia can be too much present in ways that foster dangerous desires for the purity of home and homeland.[4] Without the ameliorating factor of Hermes's opposite energies of ambiguity, communicativeness, and movement, Hestia is the hearth become a purifying inferno. "She" may demand the sacrifice of anything that would pollute the home of its historic or ancestral identity.

Therefore, Hestia reveals herself as a true inhabitant of a polytheistic cosmos; these gods do not come alone. If adhered to singly, they spell possession and disaster. With the implied presence of Hermes, Hestia offers a unique gift to the sleuth: a focused centering consciousness that is often crucial to success in the quest for truth. Indeed, James Hillman argues that Hestia is the quality of "in-ness," the structuring of psychic interiority that has been so central to the modern age.[5]

It is she who cultivates a sense of an inner self. Fortunately, she is also more than the impoverished inner being of the modern person who sees him- or herself as separate from the world. Hestia is "soul essence" that gives a quality of being at home in a body, household, city, or even the planet.[6] She gives hearth-life to all the ways we might find "home." Hestia's sense of being "in" therefore includes the hometown, the wilderness, also the sense of Earth as a planet with a core of fire as our home's hearth. In fact, Hestia converts the abysmal homelessness of space and time to the embodied, psychic sculpting of "place," the place where the hearth makes a home.

Consequently, although Hestia, mistakenly erected as a monotheism, a goddess ruling alone, may persecute those who do not belong, she is also the protector of the stranger at the hearth. She constellates in many detectives who take on cases of those outside their usual definition of home. Finally, the cozy genre, with its ubiquitous turning to the Grail Quest as a myth for meaning, has summoned Hestia to sustain the development of food-oriented mysteries.

As we will see, these catering businesses work as a mediation of Hestian powers, from the home place in the domestic household to constituting it in the home town. In doing so, Hestia is threatened by that most unhomely of crimes, murder; and so the cook becomes sleuth, in order to live fully up to her Hestian skills. Because cooking becomes a way to detect the truth inextricable from comfort and hearth-making, the food itself becomes a materialization of the Grail. Therefore, the sleuth in her quest manifests the Grail as fertility cup, sacred for bringing the goddess and her fertility as food to heal the wasteland.

But let us begin with Hestia's powers of home-making as detecting from the early country house mysteries to the making of a home through art and performance. Hestia is artful even as her creations strive to pull away from Hermes's tricky ambiguity. In the next section we will explore the contest between Hermes and Hestia for control of space. Here it is won by Hestia who makes it her place, a home.

Making Home Through Detecting

> What Mike didn't realize was that feeding the experts who worked the crime scene was definitely to her advantage. He thought he'd conned her into serving coffee and cookies, but she planned to turn the tables on him. Inviting the crime-scene experts up for refreshments would give her the chance to ask questions and to listen as they talked among themselves. Her invisible waitress trick would work beautifully.[7]

Hestia's hearth is here mobilized as a detecting and knowing organ. Hannah Swensen is resident of Lake Eden Minnesota, proprietor and cook of The Cookie Jar and indefatigable amateur sleuth of the remarkable number of unlawful killings in this rural community. She is also a trickster. Here she uses her cop boyfriend's love of her cookies to get a place at the post-crime scene informal debrief. Effectively, her food becomes a hearth around which the weary professional detectives gather. Paradoxically perhaps, Hannah here provides a home for institutions of the law, and in doing so materializes the altar upon which is deposited the known facts so far.

In the ensuing conversation during the substantial meal Hannah actually provides, the medical examiner and the cops determine what can be currently ascertained in the murder of Hannah's neighbor, the lottery winner Ernie Kusak. His outrageously noisy Christmas display has led to Hannah discovering his corpse. Although no writing is actually handed over, Hannah, as a result of becoming "the invisible waitress," does receive the facts so far. Her resemblance to Hestia's ancient altar upon which judges deposited their work each night continues when, as usually happens, her sister, Andrea, copies her police chief's husband's crime scene photos for her. In the course of her sleuthing, Hannah usually also manages to obtain the relevant autopsy report. While officially forbidden by Mike, such tricky maneuvering is effectively condoned.

So, in Joanne Fluke's food mysteries, Hestia as hearth in the form of Hannah's cooking and as repository of legal facts is acted out as both detecting and family life. It takes Hestia with Hermes skills to prevail over the institutional exclusion of Hestia powers from police work. For Hannah, single but with a life deeply entwined in family, friends, and cops as boyfriend and brothers- in-law, sleuthing and cooking are Hestian and indivisible.

Similarly, Joanna Carl's chocoholic mysteries are set in the haven of Warner Pier, where narrator Lee McKinney has retreated after a failed marriage.[8] Taking a job in her aunt's famous chocolate shop enables Lee to re-start her life, if only murders did not interrupt her chances for happiness. In *The Chocolate Bridal Bash* (2006), Lee is about to remarry when she discovers that her mother's refusal to return to Warner Pier concerns an unsolved murder. It occurred years ago, on the night that she ran away from her own wedding.[9]

Lee sleuths in order to re-knit the fabric of family life, to re-kindle Hestia's hearth. In this she differs from Hannah, whose detecting *is* the fire of the hearth. As with *The Chocolate Bridal Bash*, cozy mysteries often begin in defense of the hearth, rather than with Fluke's more comic take on detecting as familial bonding. Yet cozies thrive on Hestia manifesting, making real again, the hearth fire through the embodied, interior, and familial knowing that comes with working in/as the home.

For example, the earlier history of the cozy as clue-puzzle offers the large historic house that crime pollutes, so that it becomes uncanny. A fine example is Georgette Heyer's *Footsteps in the Dark* (1932), which begins as a haunted house mystery.[10]

Inheriting a house long shunned as spooky, three siblings find their Hestia qualities of homemaking severely tested by strange creaks and shrieks. Fortunately, a charming undercover cop enables Hestia's capacity to purify the hearth as entirely positive. Crime proves to be the source of disturbance, not supernatural threats. And yet the disguised cop first appears as the stranger who threatens Hestia, as he seems to be dangerous.

However, like many mysteries where the sleuth has to identify which is the stranger to be protected and which is the one to be expelled, Hestia triumphs. The ghostly house ends as a secure home, blessed in the form of Eros, romantically centering the cop into the new family. He is to wed the sister. It is worth looking in more detail at how family and romantic sparks constitute Hestia's hearth in detecting.

Case History (9): *A Body in the Bathhouse* (2001) by Lindsey Davis[11]

A British author whose mysteries are set in Ancient Rome and its extensive empire, Lindsey Davis's main sleuth is male, Marcus Didius Falco. He is ably supported by girlfriend, later wife, Helena Justina, a senator's daughter. Later additions to the detecting family include their dog, Nux, Helena's brothers and Falco's best friend, Petronius. Less helpful are the two daughters they produce (because small children), his mother, who is stormily estranged from his dodgy antiques dealer father, his sisters, their children and also, Falco keeps reminding us, their even more disreputable fathers. However, Helena and Falco adopt an abused girl in Londinium who grows up to be Falco's worthy successor, Flavia Albia. She has been given own series, which began in 2013 with *The Ides of April*.[12]

Falco is officially an "informer," a Roman occupation of low status that has to do with legal matters. Always beset by family, his cases are never uncomplicated by domestic complications. *A Body in the Bathhouse* is typical, in that while Falco does not want to visit Britain to investigate the extraordinary expense of a huge new palace being built for a client king, he does have a reason to kidnap his favorite sister,

Maia, to get her out of the way of a dangerous ex-suitor who happens
to be Rome's chief spy, Anacrities. A longtime enemy of Falco,
Anacrities is also sending agents to investigate possible fraud at the
building site of King Togidubnus's new home. Moreover, to cap it all,
the accident prone operation is the likely destination of the unreliable
contractors who deposited a corpse beneath the badly paved mosaic of
Falco's new bathhouse in Rome.

Humor fuels Hestia's hearth in Davis's mysteries. Romance between
Falco and Helena is both articulated through the investigations and
dedicated to securing their own domestic centering. Therefore, Helena
accompanies Falco to inclement Britain. (It is a standing joke of this
British-authored series that Britain itself is the worst place in the
empire—and the most unHestian.) Maia gets tricked onto the boat,
and Petronius, who is attracted to Maia, is left to mind her four young
children. In fact, such are the Hestian properties of Falco's investigations
that Petronius arrives with the children just as Falco has tracked the
killers of the Togidubnus's architect to a dangerous bar. They arrive at
the very moment that Anacrities's expert assassin, Perella, is about to
offer her erotic dancing to a bunch of drunk building workers.

Typically for Falco's Hestian cases, the children prove important
to the solving of a complex series of financial, ethnic, and political
crimes. Maia's sensible son, Gaius, the small daughter of the building
project clerk, and the boy who brings a hot drink each day to the
workers, all contribute clues to Falco's Hestian centering consciousness.
They even come to the rescue when nasty Brits set their fighting dogs
on Falco and Petronius.

While corruption is part of the Roman way of life, who exactly is
murderously threatening King Togi's attempt at a grandiose Hestian
hearth? As Falco would cynically interject, murder begins at home in
his world. Greed on the part of the King's trusted aide surmounts his
loyalty and leads to the unHestian elimination of the project manager.

Falco therefore works as Hestia in weaving his family life into his
detection at the same time as uncovering serious threats to the sacred
duty of the goddess to protect the hearth and the strangers who reach
it. Significantly the killer is Hestia gone dark for wanting a purity of
homeland not possible in an empire. Yet he is also guilty of corrupt
practices that lead to murder. Falco is also a sophisticated Hermes,

travelling and communicating trickily on behalf of the emperor to secure a deal for the killer that will minimize political embarrassment. It is Falco's ability to embody both Hestia and Hermes, be the centering consciousness of his family *and* negotiate shady aspects of the Roman imperium, that ensures his success as a detective.

Finally, Falco meets the last suspect in the almost forgotten incident of a dead body in his own bathhouse. Having ignored a toothache for too long, Falco visits the infamous local tooth extractor only to discover his errant bathhouse contractor when it is almost too late. *A Body in the Bathhouse* ends with a comic restoration of Hestian interiority in varieties of health. Falco gets his bad tooth removed and in turn is determined that the criminal be made responsible for his deeds. So the Hestian interior of both Falco's home and of this tiny moment within the Roman Empire is restored.

Falco always discovers that his world is more venal than his capacity to make it a real home through his detecting. In this sense the failings of his own family are metonym for a dimension of the wasteland that cannot be redeemed; an aspect of the hardboiled sleuth that Davis artfully unites with her clue-puzzle structure of the family detective. What Falco shares with all fictional detectives is integrity, that fierce quest for truth that exists in tension with their trickster hermetic qualities. Where that integrity is Hestian is in its psychological essence as a centering focusing consciousness that enables the detective to find and be at home. It is time to consider the role of cooking in producing this kind of resolution.

Hestia in Food Mysteries

Agatha Christie has some claim to have pioneered the cooking sleuth in 1957 with Lucy Eylesbarrow in *4.50 from Paddington*.[13] This enterprising young woman, like her American younger sisters, is well educated and self-motivated. Far from the historic role of the female servant encountered briefly, if at all, in pre-World War Two mysteries, Lucy runs her housekeeping and cooking services as a thriving business. She is particular about never staying long enough to be pushed into traditional feminine subservience. Therefore she is ideal to be a stand-in for frail Miss Marple, who is determined to find a corpse thrown from a train, the 4.50 from Paddington.

Unable to locate the body after the murder was witnessed by Miss Marple's elderly friend, Mrs Gillycuddy, the police dismiss the problem. So Miss Marple hires Lucy to take the temporary position of housekeeper to Crackenthorpes, in whose overgrown grounds the victim must be concealed. What makes Lucy take on Hestian qualities is the way her work in the household restores the family as the basis for letting her ask questions and snoop. Indeed, although she quickly discovers the corpse, she wants to stay because her Hestian role attracts her. Continuing to cook and sleuth becomes a way of embodying home for people she begins to care for.

> "Well as far as Miss Marple is concerned I've *done* my job, I've found the body she wanted found. But I'm still engaged by Miss Crackenthorpe, and there are two hungry boys in the house and probably some more of the family will soon be coming down after all this upset. She needs domestic help. If you go and tell her that I only took this post in order to hunt for dead bodies she'll probably throw me out. Otherwise I can get on with my job and be useful."[14]

Similarly, Isis Crawford, Joanne Fluke, Katherine Hall Page, Diane Mott Davidson, Joanna Carl, Nancy Pickard, and many more create cooking detectives whose skills are equally Hestian in the hearth-making function of the delicious food, and also Hestian in their capacity to *know* through these practices.[15] So here food materializes the healing of the wasteland and the Grail as *productive* of fertility. So it is unsurprising that food connotes romance with cooking sleuths erotically involved in detecting with cop boyfriends or husbands; for example, in the cases of Hannah Swensen or Mott Davidson's Goldy Schultz. For Hall Page's clergy wife, Faith Fairchild, her catering business is a necessary space of selfhood against the overwhelming role her marriage bestowed. Her detecting similarly shows the Hestian cooking sleuth in alliance with trickster-Hermes.

> Faith only lied to her minister husband, when it was absolutely necessary and even then crossed her fingers behind her back. Patsy's plan would make both unnecessary.[16]

Patsy's plan is for Faith's catering business to be a pretext to snoop around a Gallery where a saboteur is suspected. Faith's husband's dislike of Faith on the case differs markedly from Hannah Swenson's family and associates. They see her detecting as a dynamic part of the fabric of their home, restoring the Eden in Lake Eden. Hence a typical exchange between Hannah and mother Delores over Hannah's delicious cookies.

"Will you let me know when you find out?"

"What makes you think I am going to find out?"

Delores looked shocked. "Well you're going to investigate aren't you? You simply have to, Hannah!"

"Why do I have to, Mother?"

"Because we all have to work to catch Ronni's killer before my launch party!"

I should have known it had something to do with you, Hannah thought, but she remained silent. Verbalizing that sentiment would only hurt her mother's feelings.[17]

What connects most cooking sleuths is that by making food they put the psychic fire in the hearth of the home. Here the detective ritually enacts the distinctive interiorizing focused consciousness that is the essence of Hestia and her knowing. Cooking is a rite for connecting, ruminating, grieving, centering, *taking in to the interior of the body and becoming the body.* It is Hestia making a hearth and home even in the midst of murder and chaos.

Without realizing it, we'd all created concoctions that demanded the precise cutting of vegetables and fruit, as if organizing food could somehow order experience and make life neat. Like most folks, we believed that performing that small ritual of comfort, bringing nourishing gifts, could make life after a sudden death more bearable.[18]

Importantly, food mysteries are not necessarily conservative in their social vision. Just as Falco tends to discover Hestia banished or corrupted, so Goldy Shultz, herself a former victim of domestic abuse, is well aware that the poor are disadvantaged when it comes to crime. *Her* Hestia protection of the stranger at the hearth is often motivated by a sense of social injustice.

> She was right. I had seen it again and again. A low-income person without power is blamed for a crime and goes to jail on scanty evidence. A wealthy person, who's guilty as hell impugns the job the police are doing, impugns the victim, impugns whoever's around, and gets away with rape… or murder.[19]

It is time to look in more detail at Goldy Schultz's career as questing knight via home based catering.

Case History (10): *The Main Corpse* (1996) by Diane Mott Davidson[20]

What particularly appeals to me in this deeply realized novel is that a cooking sleuth commits an armed heist accompanied by her teenage son, and all for Hestia motives.

> Marla cackled and gasped again. "Leave it to Goldy to break me out of jail using food. Marvelous—"[21]

Goldy Schulz's best friend, Marla, is in jail because of an abundance of material evidence that she has murdered her lover, the dubious financier, Tony Royce. Goldy's admirable second husband and cop, Tom, cannot help, because the law's rigid rules of knowing privilege the material fact of Tony's blood on Marla over Goldy's intuitive, Hestia-based knowledge of Marla's heart. Her friend is really a family member, since they are both ex-wives of Goldy's first, violent husband.

After years of sisterly solidarity, and Marla's loving support through Goldy's painful healing from physical injuries, Goldy *knows* that Marla cannot be a murderer and is being framed. This form of knowing is feminine because it includes aspects of psyche marginalized *as feminine* by the dominance of separation and rationality in modernity. Hestia is one of the modes of this feminine knowing. So when conventional support of Marla fails, Goldy turns to more drastic options in Hestian family values. She draws in Marla's brother-in-law and depressed ex-soldier, General Bo Farquar. She even risks Marla's life by telling her to eat lime Jello, which will provoke an allergic reaction, dangerous to one who is a cardiac patient.

Goldy and Bo trick the speeding ambulance carrying Marla into stopping, and then kidnap her at gunpoint. By successfully treating

Marla, she survives and the fugitives escape to a cabin in the woods. Goldy's teenage son is among them, because he refuses to loan his new pet bloodhound to his mother without his presence to reassure the previously traumatized dog. Goldy and her unlikely heist team need the dog to track Tony Royce and prove Marla's innocence, which they eventually do. But first, the team needs a good meal!

> The general built a fire in the main fireplace, and soon the cabin was lit with a cozy glow.[22]

> Soon the chicken, garlic and onion were sizzling and a mouthwatering scent filled the cabin… At least I was making something for Marla that was low fat, I thought grimly.[23]

Through fire and food, Goldy as Hestia re-constitutes the hearth with the unlikely assistance of Bo. This is a fortunate development, especially as Goldy's desperation has caused her to stray into the Hestian dark side by literally *poisoning* the woman who, during the years of battering from the ex, has been both at her hearth and co-constituting it. In order to save Marla, Goldy nearly kills her, dragging her from her home in the city to a far more Artemis-like experience in the wet woods. On the other hand, the framing of Marla was so successful that Goldy's excursion into crime makes sense to her, if not to her policeman spouse.

Goldy's choice is between letting Marla, the "stranger" at her hearth who has become sacred, be destroyed, or to take her Hestia skills of making home through food into the underworld of near death. Hence, for a few scary hours Goldy, Marla, Bo, Arch, and Jake the dog are literally homeless and wandering after mounting their trick heist—perhaps a necessary enacting of Hermes—before they are able to re-light the hearth in the cabin.

One feature of this novel is the stress on relationship as a key mode of knowing in detecting, and as indigenous to the Hestian hearth fires. Goldy's involvement with people through her catering organically expands her home to the point where she even acquires aspects of Hestia as repository of writing related to law and contracts. Her assistant, McGuire, another troubled teen, decides he has a sleuth vocation and enthusiastically pursues documents that would indicate dodgy dealing by Tony. In turn these papers will provide his motive for framing Marla for his own supposed murder.

In *The Main Corpse*, Marla, an endearing recurring character in the Goldy Schultz novels, is in a Hestian nightmare. Her lover undermines her sense of home not only by framing her for murder, but also by taking her money. What saves her above all is Goldy's Hestian consciousness that mobilizes her hearth knowledge of Marla's innocence, a knowledge of "in" for both Marla and for the unknown tangle of crime and relationships that is *The Main Corpse*. When Tom tells Goldy to trust the conventions of the masculine law he says:

> "Don't get in that kitchen and start cooking and think. Oo, oo
> I'm gonna hatch something up. Please?"[24]

He is too late. Goldy is already Hestia in the underworld. Her daring and humor-fuelled crime jaunt is successful because she continues to keep the hearth for those she loves best. At the end of the novel, Goldy leads her heist team literally into the earth, entering an old mine named Eurydice to discover another murder victim and the key to Marla's problems. For a crucial part of the mystery proves to be of the underground itself. Just what does the mythically named mine contain?

In fact, it is Goldy who shepherds her team to show that false claims about the mine, about the interior of the earth, are at the heart of a criminal conspiracy. Goldy therefore explores and stabilizes the understanding of this part of her home in these Colorado mountains. In so doing she enacts the eco-Hestian home, the sense of Earth as home though its central hearth fire. Of course, the mine is not homely and like mythical Eurydice herself in the underworld, the dead person they find does not return to life.

However, at the end of *The Main Corpse* Goldy succeeds by taking others "in" to the depths of the problem. She shows that the police stress on material clues can be superficial, staying on the surface of the crime. Hestia sleuthing leads readers into relationships, the hidden depths of communities, families, businesses, and even the earth itself.

It is time to look further for Hestia's non-Hermetic relationship to writing and its pull into the interior of the mysteries of psyche and home.

Hestia's Binding Contracts

Hestia arrives in mysteries every time a Private Investigator accepts money from a client, so sealing their contract. What makes these relationships *Hestian* is the P.I.'s fidelity, which can outlive the life of the client or their attempt to rescind the bargain. Clients, like detectives, murderers, and the genre itself, can deceive. Hestian faith in the contract as binding and stable is not so mutable. Like the ancient law documents deposited on Hestia's altar, the sleuth's contractual obligations have a centering integrity to home in its widest sense of the city and community. Indeed, arguably Hestian writing enacts the notion of planet as home in its ability to center meaning. Hestia writing resists, and exists in tension with, Hermes's willful fluidity of signifying.

So in *N is for Noose* Kinsey Milhone accepts a commission to look into an apparently natural death of a Sheriff from his widow.[25] Discovering discrepancies in the story of his heart attack while out on a call, Kinsey's investigation begins to close in on the family. When trying to stop Kinsey, the widow discovers that her contract is not a personal service that is subject to her whim. Rather, once Kinsey realizes that the existing record is flawed, she *must* find out the truth, even if the client no longer wants it.

Interestingly, once Kinsey locates and learns to read the Sheriff's coded notebook, this particular truth will be deciphered. Yet it is the violent attempts to stop her detecting that ensure the code can be read, and also provides the evidence to support what was only circumstantial before. Kinsey ritually enacts the writing on the altar by ensuring that the truth is sacred and centered in the sense of being trustworthy enough to ground the community, even if by doing so it destroys the client's home.

Hestia's writing lingers in the P.I. contract, as well as appearing as the centering or grounding truth that the sleuth uncovers, even if it is not always accepted by the world at large. A detective with a political antenna such as Sara Paretsky's V.I. Warshawski often has to be both altar and writing, if what she uncovers is criminal beyond the capacity of her world to punish it. So, for example in *Toxic Shock*, while individual killers are surrendered to the cops, punishing those responsible for the corporate slaying of workers in an unsafe chemical plant cannot be realistically portrayed in this genre.[26] The best V.I. can achieve is to find another kind of Hestian altar, by handing the problem to a sympathetic lawyer.

Above all, Hestian writing means finding the right story. The right story restores Hestia's values of hearth and home, not just for the individual client or victim, but also for the human, and even non-human, community.

Amanda Cross's sleuth, Kate Fansler, is a famous college professor of literature.[27] She expresses the search for the right story as the narrative that can be a stable truth on Hestia's altar.

> "His other mistake was not to realize that I look for narratives. That's my profession, not being a detective. That's the profession of every professor of literature…"[28]

Hearth and home are not always restored by the existing authorities after the murderer has been identified. In *A Trap for Fools* an ambitious black female student was killed because she got caught up in the corruption of powerful white males. To offer some restitution to her family, and to challenge the unhomeliness of an unfair society, Kate blackmails the university provost into a new contractual obligation.

> "I thought three large scholarships," Kate said… They'll be the Arabella Jordan Fellowships, of course…"[29]

Kate's blackmail on behalf of the victim and for disadvantaged students ensures that she herself will embody Hestia's altar and the writing on it, for her price is discretion. Offering something to ameliorate the lack of homeliness at the university is a more significant centering position than merely publicizing the facts. Hestian mysteries seek the *right story* that preserves, supports, or creates the home. Hestia is here a centering energy that stabilizes meaning and does not limit it to one psyche or one person as an individual; Hestia writing is itself a hearth-making home.

Indeed, the Hestian hearth can embrace the world beyond the fictional mystery, as my argument in Chapter 1 suggested. If the solution of death is one of the mythical trickster qualities of the genre, then the solution of history is not beyond its range. Here Josephine Tey's *The Daughter of Time* is a characteristic example of attempting to re-orient the historical record.[30] The astonishing challenge of this novel is that policeman Alan Grant tries to solve murders committed five centuries ago while confined to a hospital bed.

At the mercy of William Shakespeare's partisan portrayal, King Richard III has a reputation for tyranny, compounded by the death of his two young nephews, presumed murdered on his orders. First struck by a sympathetic portrait, Alan Grant becomes convinced that the full story has not been told. When he is visited by an American researcher, Brent Carradine, Grant finds himself drawn into this historical mystery. Together they find considerable contemporary accounts of King Richard's good character and even evidence that he was loved by his people.[31] On the other hand, his successor, Henry Tudor, with more reasons to kill the two boys, is undoubtedly ruthless. In the end, Alan and Brent together make up a Hestian altar, writing for a truth not contained in history.

> Richard III had been credited with the elimination of two nephews, and his name was a synonym for evil. But Henry VII, whose "settled and considered policy" was to eliminate a whole family, was regarded as a shrewd and far-seeing monarch… Grant gave up. History was something he would never understand.[32]

Grant detects the absence of Hestia in the historical record in which history is written by the side that wins. In effect, what Grant and Brent do is to accept Richard as the stranger at their unlikely and temporary hearth of Grant's hospital room. Sensing his innocence from the unreliable signs left by time, Grant intuits, and Brent researches, to provide an instance of Hestia centering as the quest for the right story. Together they provide a sense of home in which even after centuries gross injustices can be addressed. If King Richard can be entertained at the modern hearth of sleuthing, then what about more immediately vulnerable strangers?

Protecting the Stranger at the Hearth

Not all cozies have a Hestia-like embrace of the stranger at the hearth. For example, corpses in Joanne Fulke's Lake Eden tend to represent a disruption to the close knit community, such as "lunatic Larry Jaeger's Crazy Elf's Christmas Tree Lot."[33] Even more indicative is the threat posed by Bev, ex-fiancé of Hannah Swenson's sensible suitor, Norman. This devious dentist insinuates herself back into his life.[34] By claiming her daughter is actually fathered by Norman, Bev persuades him to set a date for their marriage, so provoking Hannah to mobilize her detecting abilities on his behalf.

Of course, as Bev proves treacherous, Hannah's extended family of suitors, as well as friends, mother, and sisters is reconstituted. Hannah is a Hestia expelling pollution from the home, although she also welcomes employee, the redoubtable storyteller Lisa, to her hearth. More typically Hestian and in a way that spans the cozy-hardboiled wide spectrum, is the stranger who is sacred because client (for the P.I. sleuth) or victim on whose behalf Hestia must renew the wasteland. Those sleuths inhabiting an irredeemably corrupt society, such as Sue Grafton's Kinsey Milhone, and Lindsey Davis's Falco, are doomed to discover that the truth is not enough.

In Grafton's *K is for Killer* and *N is for Noose* the client learns the true extent of Hestia dysfunction in their own families. While in Davis' *A Dying Light in Corduba*[35] and *Time to Depart*,[36] Falco learns, not to his surprise, that unjust politics and organized crime will persist beyond his identification of murderers. Indeed, Falco begins his career in *The Silver Pigs* when a young girl he likes is murdered and he, Hestia-like, takes on finding her killer as a sacred trust. The truth does renew a family, but does so by the rather drastic method of the violent death of another member, whose body Falco must conceal for political reasons. However, the prime domestic suspect, Helena Justina, becomes the love of Falco's life. Aphrodite facilitates Hestia as they make a new family.

Several authors specialize in consecrating the stranger as the one on whose behalf the dangerous quest for truth is undertaken. Both Annie Darling for Carolyn Hart, and Goldy Schulz for Diane Mott Davidson, adopt vulnerable people who are either murdered or who are too easily suspected of it. Another Carolyn Hart sleuth, Henrie O, an ex-journalist, also gets drawn into detecting by her Hestian warmth. In *A Scandal in Fairhaven* (1994), Henrie is staying at a close friend's house when a young man breaks in covered in blood.[37] He is subsequently arrested for the murder of his wife, and the police seem uninterested in considering the possibility of other suspects. Partly out of obligations to her friend, kindly Henrie is drawn in to help him.

Unlike Annie Darling, who remains wedded to her Carolina island, Browards Rock, as well as romantic, rich husband Max, Henrie O travels and yet still finds herself embedded in Hestia work in re-making those homes and families she feels personally connected to. Another stay-at-home sleuth is British writer Jill Paton Walsh's Imogen Quy, a college nurse at a fictional Cambridge College.[38] In *The Bad Quarto* (2007), not only does her lodger, Fran, motivate her to detect when her drama society gets caught up in murder, but the mysteries themselves are based on the extent to which Imogen can *be at home and make a home for others* in the prestigious college where she works.[39] So Imogen comforts the vulnerable young students, and is called upon by the genial college Master when threats to its domesticity appear, such as problems with a wealthy donor, in *Debts of Dishonour* (2006).[40]

Mary Daheim has a whole series devoted to the Hestian phenomena of the stranger at the hearth as sacred or threatening. Her bed and breakfast mysteries feature Judith McGonnigle Flynn as the manager of an establishment beset with expiring guests. Fortunately she is married to an ex-cop, Mike, and has an enterprising cousin, Renie, to call upon. Family life in the Seattle B & B is a constant negotiation between who is manipulating this lively domestic setting for criminal purposes and who is the stranger to be protected? Or, if a victim, the dead guest is the source of the sacred charge to find the right story.

In *Snow Place to Die* (1998), Renie persuades Judith to cater for a corporate retreat, only to find that they get snowed in with the fractious executives.[41] Once corpses appear, this updated version of the traditional country house mystery becomes a constant sifting to find out who is destroying and who trying to nurture the distorted home that is this high powered business. With the last victim the "home-maker" of the company and the CEO's longtime mistress, the corporate family implodes. Only Judith and Renie can reassert Hestia values of care for this once functioning community. By detecting the killer as the patriarch who has lost control of his business-family, Judith and Renie are able to protect the remaining strangers until the police can be convinced.

Now it is worthwhile looking at an example of Hestia where the sleuth is seriously lacking in the domestic arts.

Case History (11): *Seven Up* (2001) by Janet Evanovich[42]

> In my mind my kitchen is filled with crackers and
> cheese, roast chicken leftovers, farm fresh eggs and
> coffee beans ready to grind. The reality is that I keep
> my Smith & Wesson in the cookie jar, my Oreos in
> the microwave... and I have beer and olives in the
> refrigerator. I used to have a birthday cake in the freezer
> for emergencies, but I ate it.[43]

Stephanie Plum is a bounty hunter in Trenton, New Jersey and
narrates a series of sleuthing adventures where comedy defines the style
and tone. Like Hannah Swenson in Lake Eden, Stephanie has two
suitors. Unlike Hannah, sex comes into her detecting. Stephanie is
involved with a cop, Joe Morelli, who is a fantastic lover and also,
infrequently, with Ranger, a security operative who works outside the
legal system and whose sexual skills are enough to make women faint
on sight. Stephanie also has to negotiate her parents' home, a source of
wholesome food and Grandma Mazur, a gun-toting senior with a roving
eye and adventurous spirit.

Also like Hannah in Lake Eden, Stephanie's district, "the Burg," has a
strong Hestian sense of community as home, if somewhat more fragile.

> It's really a very safe neighborhood, since Burg criminals are always
> careful to do their crimes elsewhere. Well, okay, Jimmy Curtains
> once walked Two Toes Garibaldi out of his house in his pajamas
> and drove him to the local landfill... but still, the actual
> whacking didn't take place in the Burg. And the guys they found
> buried in the basement of the candy store on Ferris Street weren't
> from the Burg, so you can't really count them as a statistic.[44]

Unlikely as it seems, Hestia connects the almost idyll of Lake Eden
to the hardboiled wasteland of Evanovich's Burg. Both places provide
a centering sense of security where generations of inter-related families
respect unwritten rules. These serve to anchor psyche in family and
community. Where the Burg substantially differs, of course, is in its
proximity to the underworld of "Family" as organized crime. Here
Hestia is indeed partnered with Hermes in ways that both knit together
and rive apart those families interconnected genetically and those bound
together by lives of crime.

As trickster and messenger, Hermes darkly facilitates revenge when aged Mob member Eddie de Chooch, already wanted for failure to appear in court, mishears an order and removes the heart from the body of a dead Mafioso. With the widow demanding that, on pain of death, the heart be restored before burial, Eddie has plenty of motives to resist Stephanie's attempt to cuff him for a more minor offence. After nearly blind Eddie gives Stephanie the slip, she discovers a corpse in the garage of supposedly blameless Burg resident, Loretta. Only Stephanie's and the Burg's sense of Hestian loyalties will enable her to offer a heart (actually from a pig) to the crazed widow who has kidnapped two of her friends, save intrepid Grandma Mazur when briefly snatched by Eddie, avoid Ranger's exciting overtures until the last moments of the novel, and even manage to witness her "perfect" sister's return to the family home after her marriage collapses.

Yet if Hermes has a demonic side in *Seven Up* in fostering criminal trickery and doubling "family values" (as both mafia code and the equally rooted Burg respectability), we also see how Hermes facilitates a vital flexibility in Hestia, enabling her to be necessary familial centering in organized crime *and* the working class values of the Burg. Stephanie is stalked by Ziggy and Benny, who are also looking for wandering De Chooch. "Associates" and friends of fifty years, Stephanie is never sure if they are more motivated by nurturing familial concern or the needs of the "Family" to have deadly secrets kept. In a sense, when linked to Hermes, Hestia's presence is vital (in the sense of sustaining a hearth that promotes life) to the ability of the two types of family to co-exist in the Burg. Given that respectable cops, like Morelli, have relatives in the Mob, it takes a trickster to negotiate family loyalties when they come up against the codes of the Family.

Moreover, Hestia without Hermes, without flexibility and communications beyond the hearth, can also be centering in hell. The fearsome widow of the literally heart-less corpse tortures hapless Moon and Dougie, stoner friends of Stephanie, demanding "purity" via the intactness of her husband's body. She is prepared to kill for Hestia. Not only do her warped family values make for hell, in her complete severing from the community's ability to juggle, trickster-like, the competing Hestian codes, but she has been driven murderously crazy by confinement in her home. Both types of family values, shame over a schizophrenic wife and a desire to keep it "in the family," mean that her husband created a prison cell for her in their home.

Stephanie too learns the necessity for Hestian powers of home to be linked to Hermes' trickster ambiguity. Desperately dodging her mother and grandmother's marriage plot for her and Morelli, she is encouraged by seeing that Valerie is welcomed home, even in the midst of the failure of her own. And although her parents' Hestian devotion may be stifling, Stephanie is at last able to acknowledge her need to worship Hestia by darting between her own, sometimes over busy apartment (where lovers and criminals seem to break in at will), and her ancestral hearth. This, we learn, is a Hermes-with-Hestia centering from childhood, when Stephanie would frequently run away to Grandma Mazur's house.

Without Hestia, the Burg would be a criminal wasteland; even though she also enables crime families to center their activities in an enduring hearth (so proving her psychic kin to Hermes). Without Hermes, Stephanie could not negotiate the dual modes of family in her job, nor maintain her primary Hestian allegiance to her parents' house and the Burg as hearth and home.

Perhaps it is time to consider Hestia's absence more deeply as a motivating fact for the sleuth and for the mystery genre itself.

Detecting Hestia Absent

As this chapter has so far shown, mystery fiction is a tricky genre in which any affinity for dualism is accompanied by dismemberment of simple binaries (see Chapters 1 and 2). So, it is not the case that Hestia simply in-forms the modern cozy while the hardboiled P.I.s disown her. One way of understanding Hestia's mystery role historically is to consider the contemporary cozy's ancestor, the 1920–40s clue-puzzle, as Stephen Knight terms it.[45] Although on the one hand Hestia might appear native to these fictions in their predilection for domestic murders and country house settings, in fact the historic clue-puzzle more typically detects Hestia's absence.

Four British women authors, often called the Queens of Crime, led the development of the clue-puzzle in the inter-world war years: Agatha Christie, Dorothy L. Sayers, Margery Allingham, and Ngaio Marsh.[46] All of them set their opening murders in familial settings in which the sleuth is a stranger. Hestia occurs as the stranger at the hearth as sacred in the divine knowledge reached. The detecting is divine because it does save members of the household from a wasteland that results from suspicion. No one but a special, Grail knight sleuth can solve or dis-solve the disintegrating taint of murder at the hearth. Yet this wasteland is not sourced in the murder but rather constellated by it. All these mysteries begin in the revelation of Hestia absent in the households beset by jealousy, greed, or insane obsession; it is these which are destroying the Hestia consciousness that makes a home.

Christie begins with *The Mysterious Affair at Styles* (1920) in which Captain Hastings, convalescent from the First World War, is invited to Styles Court by friend John Cavendish.[47] There he finds suspicion surrounding the second marriage of Cavendish's mother to uncongenial Alfred Inglethorpe. When Mrs. Inglethorpe is poisoned, Hastings calls in a retired Belgian police detective, Hercules Poirot, to investigate. Poirot succeeds because of his uncanny ability to penetrate domestic codes and rituals while disarming suspects with his fanciful foreign manner.

Indeed, Poirot is to make a career of being Agatha Christie's gentle satire of British suspicion of their French neighbors. As a *Belgian,* Poirot belongs rather to an historic ally, even though his friend's name of "Hastings" is the place where the invading French last conquered the English in 1066 AD. Poirot and Hastings become a team signifying English ambivalence about home and neighbors. Typically, Poirot's nose for domestic details as important clues detects where Hestia is being falsified to mimic family life. Hence it is the detective who comes to re-embody Hestia for the stricken family. Only by constituting the solution as the reigniting of a family hearth can *The Mysterious Affair at Styles* offer a promise of ongoing familial bonds.

While Christie, Allingham, and Marsh all begin in the iconic *country* house, Sayers sets *Whose Body* (1923) in London, where the tight proximity of houses and families is the key to solving the crime.[48] Architect Mr. Thipps channels enough Hestian warmth into his profession that when he suffers her absence by discovering a dead body in his bath, a client calls the Dowager Duchess of Denver. She in turn calls her second son, amateur sleuth Lord Peter Wimsey, to come to his aid. The dead body is a naked man wearing only a pinze nez. Lord

Peter soon links him to the disappearance of a prominent Jewish businessman, and yet the body is not his. It takes Lord Peter's familial contacts with the missing man's relatives, as well as his talent for discerning long jealousies and professional obsessions, to uncover the real story.

Margery Allingham's *The Crime at Black Dudley* (1929) introduces her longtime detective-adventurer, Albert Campion, who shares Lord Peter's insouciance.[49] On the other hand, in not assuming a detecting role until late in the story, Campion is unique amongst the sleuths of these four early novels. In fact, Campion begins as a major suspect in a country house mystery in which a cursed jeweled dagger becomes a murder weapon. He proves first to be the vulnerable stranger as he is being sought by a gang of violent criminals.

So, *The Crime at Black Dudley* offers Hestia disintegrated, since this hearth is cold and no one knows who can be trusted to find the truth and re-ignite it. Yet Allingham's novel shares a typical plot motif with the more ordered world of Ngaio Marsh, in whose *A Man Lay Dead* (1934) a murder *game* is used to do the actual deed.[50] Murder games, still popular today, rely upon a cultural knowing of precisely this type of country house clue-puzzle. A group of guests stay in one large house and one of them is secretly given the task of murdering another (fictionally) in a way that will challenge the sleuthing abilities of the rest. Of course, if one happens to be playing the game with a rich relative who is denying vital funds, then perhaps a real opportunity presents itself!

The country house murder game as an actual murder opportunity is a perfect demonstration of the self-conscious, self-referential nature of detective fiction. It is the trickster in the genre. It is also a revelation of Hestia's absence when she is supposed to be present. Family and friends gather at the hearth only to find no sacred protection, but instead desecration by murder. In *A Man Lay Dead,* wealthy selfish Charles Rankin gathers together jealous lovers and those who will benefit financially by his demise. One of these is young Nigel Bathgate, a journalist who persuades investigating policeman Roderick Alleyn to allow him to sit in on interviews with suspects.

Alleyn is no ordinary policeman. Son of a well-connected diplomat, like Peter Wimsey, he shares the class of the privileged suspects in most of his mysteries. Also like Wimsey, he is attractive to women and prepared to use erotic charm to get results. Investigating largely by one to one questioning, Alleyn uncovers Hestia's absence and begins to use his Aphrodite affinities as fuel. Through a ritual of focused questions, what is hidden is revealed, which is not unlike the analytic frame of psychotherapy.

Alleyn's detecting interviews thereby unite Aphrodite and Hestia to re-kindle the hearth through solving the murder that is laying waste to the household. Indeed, there is often something a bit too cozy in Alleyn's welcome into the suspect's home as one of their own class. Given that he deliberately makes use of his Hestian appeal to suspects, thereby tricking them into unguarded revelations, there is something of Hestia's necessary partnership with mobile (in every sense) Hermes. The question of how far a Hestian sleuth requires an alliance with Hermes will be examined in the next section.

Hestia Needs Hermes

Jacqueline Winspear's recent mysteries with detective/psychologist Maisie Dobbs are set in the clue-puzzle environment of 1930s London and surrounding countryside.[51] After her mother dies, Maisie is hired as a servant in the aristocratic household of kindly Lord and Lady Rowan. While the serving class are largely invisible in interwar clue-puzzle mysteries, Maisie's unusual aptitude for learning is noticed when the Rowans discover her reading books in their library. Their friend, the enigmatic and highly connected Maurice, takes on Maisie as a pupil. Eventually she goes to university, until the First World War interrupts her studies. She feels impelled to serve as a nurse near enough to the trenches to be wounded along with her doctor fiancé.

The first mystery, *Maisie Dobbs* (2003), begins in the post-war period with Maisie's fledgling detective agency staffed by herself and wounded ex-serviceman, Billy. Maisie's career represents a struggle over Hestia subtly evoked in this fascinating novel of war-devastated lives. She had been removed from a happy family home to live as a servant, who by definition is not a full member of the Hestian hearth. Yet, tacitly partly adopted by the Rowans, Maisie found a second home in their world, while not forgetting her working class origins. In her gift for domesticity she embodies a fortunate coalition of Hestia and Hermes with the tricky god enabling her to be "at home" in two worlds. Maisie demonstrates Hestian loyalty by financially and psychologically supporting the wounded Billy and his family, while possessing the mobility of a Hermes in lightly moving between disparate sections of a stratified society.

Above all, Maisie is well equipped to detect how Hestia may require her alliance with Hermes in a world in which war has wrecked the very capacity to be *at home*. Investigations into mysterious deaths of ex-servicemen center on "The Retreat," a so-called closed community of traumatized veterans. Unfortunately, Billy and Maisie discover it to be the Hestian hearth at its most dark. Managed by a man himself traumatized by being forced to shoot apparent deserters in the war, men who were too shell-shocked to obey orders, this man now re-enacts executions on anyone who plans to desert this particular "hearth."

In fact, in this and later Maisie Dobbs novels war trauma proves the essential connection between Hestia and Hermes. Without the mobility of psyche, communication and circulation of new possibilities, without some of Hermes trickery, Hestia can be a defense of home, family, and self that locks in pain in the cause of a purity worth killing for. Hence, we see internecine murders in which the aim is to evict the stranger who threatens the class, racial, or familial integrity of the hearth.

Such killers can be found in the clue-puzzle country house mysteries such as Ngaio Marsh's post World War Two *Final Curtain* (1947), in which a mother schemes to protect the inheritance of her son from an unworthy gold digger.[52] Yet even in Marcia Muller's later hardboiled *Listen to the Silence* (2000), detective Sharon McCone finds a man who kills in an attempt to stop his family from being "polluted" by the child of his white son and a Native American woman.[53] In fact, here Hestia is rescued by Hermes. This racist killer is tricked, enabling the mother to live and the child, Sharon herself, to be adopted by a distant relative.

Also Hestia and Hermes become allies in women's mysteries when the Hestian-inclined sleuth faces the challenge of the possibility that a member of her own hearth is involved in crime. Can the Hestian sleuth detect the truth if it implicates someone in her home, whether that home is her own family or the extended family of a close-knit community? Some authors solve this problem by managing to locate suspects as "outside" the hearth, such as the unlikeable outsider killers in Joanne Fluke's Lake Eden. Others have their sleuths employ a Hermes-like flexibility in renegotiating their sense of family and belonging, such as with Laura Lippman's apprentice P.I., Tess Monaghan. In *The Sugar House* (2001), Monaghan needs to reconfigure Hestia in order to deal with her father's complicity with a corrupt politician.[54]

When crime threatens within the orbit of the hearth, Hestia in a positive alliance with Hermes can enable a flexible, even tricky, renegotiation of domestic stability. Tess acquires sufficient *herme*neutic interpretative skills to detect her father's essential love and fidelity to his family. Similarly, back in Ancient Rome, Falco's irrepressible and irresponsible father persists in trading on the very fringes of the law, so attracting some seriously unpleasant associates. Fortunately, Falco manages to maintain both a sense of Hestian stability in his home while also grudgingly nurturing a mutual love with his tricky father.

Hermes is truly a required ally in maintaining a vital, *vitality-producing* flexibility in Hestian focused consciousness. Again this core attribute of the goddess proves invaluable to the fictional sleuth.

Hestia as Focused Centering Consciousness

> "[E]ven though a yard official is supposed to have no psyche I find there is often a moment in a case when a piece of one's mind, one's feeling, one's sense, knows the end while all the rest of the trained brain cuts this intuitive bit dead."
>
> —Ngaio Marsh, *A Man Lay Dead,* 1934

Hestia fosters intuition as a focused centering consciousness that is the psychic manifestation of hearth and home. As James Hillman explains, Hestia is the consciousness of "in" which includes being "in" the body, "in" the family, the home, the community or "in" a place as home.[55] Roderick Alleyn's habitual mode of detecting is to arrive as a stranger at the hearth amongst a family or theatre community in disarray around a corpse. By his empathy and charming upper class ways he makes himself sacred by incarnating an intuition, a consciousness of "in" this group that detects the killer and restores its primal hearth.

A surprising range of fictional sleuths by women writers use Hestian consciousness as a necessary mode of knowing. Perhaps it is to be expected that cooking sleuths such as Goldy Schultz for Diane Mott Davidson, Hannah Swenson for Joanne Fluke, or Libby for Isis Crawford, all find preparing food a practice of psychic centering leading to insight about the cases. However the female hardboiled detectives also make use of Hestian centering, such as Sue Grafton's Kinsey Milhone who uses cards upon which she puts key facts of the crime. Typically in the novel there comes a time when Kinsey sits down, spreads the cards out, and lets her mind wander around the material until it can find a home, or essence, in some leap of knowing.

Yet it is Marcia Muller's Sharon McCone, sleuth and detective agency owner, who offers the quintessential example of needing to find Hestia in herself in order to solve the most intimate case of all, the mystery of her own origins. To which hearth does she belong? In *Listen to the Silence* (2000), Sharon discovers papers left by her newly deceased father that inform her that she was adopted. This shattering discovery is compounded when none of her (now adoptive) family will tell her what happened to bring her to them. Forced to detect amongst her closest relatives, her partner and professional hostage negotiator, Hy Ripinsky, tells her that only a special kind of listening and focusing will help.

> "You've known these people all your life. You're tuned in to their personalities, their ways of thinking, subtle nuances. Tune out their words and listen to what's hidden in the spaces between them. To the pauses, the hesitations. Picture them at the times when they won't look you in the eye."
>
> "Interesting approach."
>
> "You try it. Listen to the silence. It can tell you everything."[56]

By evoking a centering consciousness of what is behind the words of those she loves, Sharon tracks down the Native American world of her birth family. When she is told that here "family" signifies not so much those bonded by blood ties as those chosen to be trusted, she learns another Hestian lesson of the sacred hearth. She also learns that much of her instinctual ways of dealing with people are indeed "native" to her birth people. It is a deep interior struggle, but *Listen to the Silence* ends with Sharon having affirmed both the home of upbringing and a real connection to her birth family.

Practicing Hestian consciousness enables Sharon to move from feeling suddenly strange at her habitual hearth to occupying it as sacred through the knowledge she has acquired. Detecting the mystery of her birth becomes hearth-knowing because she discovers a truth of enduring bonds that re-kindles her adoptive hearth. Now she also becomes the sacred stranger at the hearth of her birth mother. She is a stranger become sacred because what she discovers protects her birth family from a man whose fanaticism over racial purity incarnates a dark deadly Hestian impulse. Sharon's sleuthing enables her birth mother's hearth to welcome her as the stranger who is sacred because she brings Hermes and Hestia back together in a powerful blend of detecting skills.

Indeed, Hermes and Hestia unite in the focused consciousness recommended by Hy. Here most overtly, must the detective use Hermes as hermeneutics in deciphering the tricky texts presented by relatives. Yet Sharon uses her Hestian consciousness of being "in" the McCone family *and also* Hermetic skills of interpretation. The result is to re-ignite the hearth in sacred knowing that is Hestia because trusting of love.

> I put my hand on hers where it lay on the table. "Ma, the family's intact. Elwood feels like a father to me, and that's good because I miss Pa a lot. But Kia will be more like a friend or a favorite relative. I already have a mother."[57]

Hestia is a powerful detective of familial qualities of home. Moreover additionally, as we will see, she has an ecological dimension, appearing when we are at home in a place. She is our planet as home.

Hestia for the Home-Place, for Earth as Home

So far this chapter has argued that Hestia is a goddess of hearth and home who can vitally drive a detecting story. The quest for truth indigenous to mysteries can also be the need to make, restore, repair, or re-found a home, family, or community. Finding the truth becomes necessary to the sleuth's sense of being at home in her own psyche. Hence a very real dimension of the sacred Hestian hearth is *place* as opposed to space, being as embodiment *within* a particular natural and cultural environment.

Here detective fiction by women has a number of strategies, such as re-inscribing or revisiting the same place in a mystery series set in a specific location. Such is the case in Mary Daheim's Alpine books, set in a North Western town called Alpine; Joanne Fluke's Lake Eden; Agatha Christie's St. Mary Mead; and Carolyn Hart's "Death on Demand" books set on the island of Broward's Rock.[58] Those places that become home via detecting are not limited to rural idylls, although cities such as Sara Paretsky's Chicago and Laura Lippman's Baltimore have to work harder to construct a fragile Hestian hearth alongside urban unease.

Indeed, one differentiation between the cozy and hardboiled genres might be that the former sleuth restores or even finds a home through detecting, while the hardboiled detective discovers just how unhomely modernity can be. She (or he in such sleuths as Lindsey Davis's Falco), finds that the truth is insufficient to heal the wasteland and make it bear reliable fruit in homeliness. The fragile hearth of the hardboiled detective is that home of family and friends renewed through the detective process; a process which is often deeply challenged by the revelations of endemic corruption. For example, in ancient London Falco uncovers a persistent parasite of organized crime in the city. That he and Helena rescue one abused child, who becomes their foster daughter, Albia, is a metonymic testimony to their Hestian relationship.[59]

It goes without saying that cozy mysteries tend to be located in places pleasing to the eye, with an emphasis on rural as opposed to urban landscapes. Famously residing in the English village of St. Mary Mead, Agatha Christie's Miss Marple triumphs in her first full-length mystery iconically named *The Murder at the Vicarage* (1930). Both book title and feminine village name suggest a religious dialogue in the sacred origins of the detecting myth. Miss Marple's investigation reconstitutes the Grail that is the integrated community. She does so in ways honoring Hestia as transforming disembodied space into place: the hearth as community home.

For Hestia's ecological sense of the planet itself as hearth it is worth pausing on the high proportion of bad weather in some cooking cozies. Inconvenient snow frequently disrupts both Goldy in Colorado and Hannah in Minnesota as they attempt to drive delicious food to ungrateful clients. Extreme heat is also a problem in both places, while Faith Fairchild in New England shares the biting winters. Of course this reflects the realistic aspect of the genre and the American climate;

it also recalls both the wasteland aspect of the Grail myth and Hestia as goddess of our ecological hearth.

In the Grail myth the sacred vessel is a cup of nourishment as the land is restored to fertility. Hestia is the hearth fires within the planet as well as the community. So why is nature in cozies often so unhomely? What makes these cooking mysteries particularly enact ecological as well as social Hestia is the way the story makes a nourishing hearth while facing inclement wild nature, one that appears not to offer a safe home for anyone. In summer heat, Goldy nearly gets incinerated by wildfires, while in winter she stumbles across bodies in blizzard conditions. One such as the snow-covered corpse in *The Cereal Murders* (1994) at her son's school, Elk Park Prep.[60] Yet it is the "bridezilla," Billie, who in *Fatally Flaky* (2009) gives a clue to the unusual rate of tempests in cooking mysteries.[61]

> Still, three weeks of unremitting, incessant downpour was uncharacteristic. The New Age people would have said that Billie Attenborough's nutty behavior had brought on the bad weather. When I told Tom that interpretation, he pulled me in for a hug and whispered, "At least we know who to blame."[62]

Mysteries are about finding out who to blame. While in a realistic sense, bad people do not cause bad weather, in a mythical sense they do. Hestia as ecological hearth is about finding the center, restoring the balance that makes a home of the planet. Metonymically expressed in the cooking, Hestia divine energy restores nature by enabling it to support community and family. Hestia is not of the wilderness as Artemis is. She takes the unpredictable and uncontrollable fire of nature and brings it to the hearth. Hence, the unpredictable and hearth-inimical qualities of human beings that lead to murder can manifest in nature and weather in these cooking mysteries.

In that solving crimes does not tame weather, the genre here preserves its tangential relationship with realism. However, Hestia herself signifies a different sense of nature—that human arts, such as homemaking, are part of its planetary creativity and not separate. Hestia is the biosphere as nurturing community. So in this sense, violent and dangerous weather expresses the hearth-destroying malice that culminates in killing. To the Greeks to murder someone who has been at your hearth was to violate a sacred place. Storms, snow, and wildfires all enact equally the murder and the consequent wrath of Hestia, as the hearth is desecrated.

Similarly, harmony in nature comes when the hearth is restored by the re-storying of the community through the actions of the cooking detective. One moment like this occurs at the end of *The Main Corpse*, when the ashes of a loved and troubled man are taken into the wilderness.

> Arch shook his head, then squinted at the trees.
>
> "Mom," he said softly. "Everybody. Look."
>
> We turned. Moving through the sunlit trees was a solitary wisp of vapor. It seemed to have a military bearing.[63]

Cooking cozies are about the murder of someone at the hearth and its necessary re-making through forms of loving knowing made substantial in preparing food. In this way, Hestia makes the wild into home by solving, dis-solving that which destroys the hearth.

If Hestia is a goddess who breaches modernity's division between human and nature, does that make the domestic detective into someone who *connects us* to the cosmos? Ann Granger's Cotswold set mysteries feature a British policeman, Alan Markby, and his slow courtship of a newcomer, Meredith Mitchell. Concluding *That Way Murder Lies* (2004), a heron returns at sunset.[64]

> Against the darkening sky a shape appeared, nearing the spot. With a rush of wings it swooped down and landed close to the jetty. Spike settled his feathers, squawked once triumphantly, and slowly and with dignity began to patrol his domain.[65]

Named Spike by humans, the heron's return to *his* domain signals a harmony between culture and nature, as the bird reclaims a home that includes the jetty. Here too sleuthing success is answered by the place as home, ecologically as well as domestically.

Finally, a particular and distinctive presence of Hestia in women's mysteries is to renew place in the cultural history of such complex territories as the United States. An example here would be "Southern cozies" which ally Hestian sacred hearth-making to an exploration of the place of cultural difference within the possible home of the nation. Toni L. P. Kelner's Laura Fleming mysteries span social divisions in this subgenre by being focused through a sleuth who has *been away*, and so can negotiate perspectives other than the possibly extreme Hestian sensibilities of her home town.[66]

In *Down Home Murder* (1993), here too Hestia needs the balance of Hermes to root out racism within the extended white clan of Laura's family. Providing the hermetic vision that ruptures Hestian dark tenacity of purity and stasis, Laura feels estranged from her family, who stubbornly persist in putting clan loyalty above the sleuth's essential quest for truth. Only with a real struggle is Laura able to re-found her familial hearth in relation to her hermeneutics of detecting. Her home town is at last re-made by detecting her home as place.

> Still, mostly what I felt was relief that despite everything that had happened, I was still family. Or, maybe I was finally family.[67]

Case History (12): *Southern Discomfort* (1993) by Margaret Maron[69]

In Margaret Maron's Deborah Knott mysteries, the southern girl returns as a lawyer and becomes a judge, despite her family's long running extra-legal business of bootleg whisky![68]

The opening of this North Carolina set novel juxtaposes the "old adversaries" of mockingbird and cat, with the former distracted by feeding babies. Far more traumatic are the relationships of human parents and their offspring with the men abusing their daughters. Here the sacred hearth is dark with incestuous and forbidden rites. Then there is the mysterious purchase of an elderly pesticide containing a high proportion of arsenic.

Yet Deborah Knott is happily ignorant of domestic abuse while celebrating her unexpected promotion to a Judge, an election owing everything to family ties!

> To everyone's surprise – everyone who did not know about the keg of Republican dynamite my daddy was sitting on – our Republican Governor appointed me.[70]

Much of the plot centers on the building of a house for a homeless family by a women's group. Beneath the camaraderie of women building without the skeptical assistance of men, a more sinister form of sisterly solidarity emerges. A father suspected of abusing his daughter is poisoned, making way for a revelation of similar crimes. Poison victim Herman survives to be defended vehemently by sister, Judge Deborah, from accusations that his relationship with teenage daughter, Annie Sue, is anything but a father's reluctance to see his daughter grow up. Unfortunately, Annie Sue has two friends who both have deeply troubled families, one of whom misunderstands Annie Sue's melodramatic outbursts against parental tyranny.

Southern Discomfort reveals the tension between hermetic qualities of change, such as creative fluidity in gender roles, and Hestian defenses of the purity of the hearth as home and community. Family loyalties drive politics and challenge the even-handed operation of the law; for example, when Deborah exclaims to the Sheriff that her sister-in-law is no killer. Deborah, in fierce Hestian mode refuses to be a disinterested officer of the law. Eventually she discovers that the murders were committed by a young girl, who is arguably attempting to restore Hestia as guardian of home from its dark perversion of sexual abuse and attempted rape. For this reason she may escape the rigors of the law, for mercy, therapy, and rehabilitation.

Hence Deborah detects Hestia as an infernal hearth in domestic abuse. Appropriately for a judge sworn in by her father handing over her mother's bible, she restores a sense of home to the community by fostering a recuperative approach to the killer. So this sleuth restores Hestia to psychic balance by tracking down the underworld she can erect if too exclusively worshipped. Hestia needs to be in a creative tension with hermetic "give" or fluidity that makes family flexible enough to survive and psychically thrive. Thus partnered with Hermes, Deborah is working for Hestia in making her family network subtle enough to maintain the hearth as home to tricky teenagers.

It takes the trickster imagination of Hermes to see that Hestian privacy necessary to family life may also disguise Hestian decay into torture and incest. Deborah catches the killer as Hermes; she tricks the culprit into trying to steal a video tape of incriminating evidence. In this way she discovers that her judgeship was owed to the murder of a child abuser, as well as to her father's blackmail on her behalf. Hestian at core, Deborah does not resign her position, but resolves to accommodate her legal status to the best of the home values of her family and community.

Southern Discomfort is a complex negotiation of sometimes contrary sacred energies in which Hestian community as home predominates. For Hestia presides in those mysteries in which a sense of home defines a place. Such mysteries summon Hermes to supply his trickster skills in order to "get round" the law's rigors in favor of preserving and renewing the embodied hearth as home. So, when it comes to her father's not necessarily finished career with illegal spirits, Deborah presides over a Hestian courtroom tempered with Hermetic trickiness. Deborah Knott makes Hestia a legal spirit in her judgments, in which writing possesses stable sacred values of fidelity to truth. Yet she also knows that her Hestian presence in court owes a lot to her Hermes father; she also knows that signs and writing can be tricky.

Epilogue: Hestia Feasting

> Weddings, funerals, christenings—most solemn ceremonies are followed by food and fellowship, and a swearing in is no different. Once all the official documents were signed, we followed the crowd downstairs...
>
> —Margaret Maron, *Southern Discomfort*, p. 13

Hestia gives humans feasting in the sense of sharing a sacred bond through preparing, serving, and eating that which nourishes and makes our bodies into a home. Hestia turns harvest (the goddess, Demeter) into home-making. Yet, in a polytheistic psyche, when Hestia is too long solitary the Hestian community is plagued with her underworld aspect. Families breed murder as well as a focused consciousness fed by hearth fires of love. A killer from a dysfunctional Hestian hell stalks Deborah Knott's swearing in as Judge, just as a family meal between cousins is a means to murder in Dorothy L. Sayers's *Strong Poison* (1930).[71]

Fortunately Lord Peter Wimsey has an expanded hearth where the flexibility and trickery of Hermes enables him to accommodate Bunter, his "man;" a retired burglar; a spinster who is a shrewd sleuth herself; a chief inspector of police; and the best defense lawyer in England. He also has the ability to befriend the woman he loves who faces death by hanging for a crime she did not commit. No wonder the accused woman feels able to envision Hestia as the debonair detective leaves her prison cell.

> "I will give the footman orders to admit you," said the prisoner gravely; "you will always find me at home."[72]

Hestia can offer feasting even in the prison cell of an innocent woman sentenced to hang. To get her out, the sleuth will also invoke the aid of her sister goddesses. For this reason, it is time to go hunting with Artemis.

CHAPTER FOUR

HUNTING WITH ARTEMIS

Introduction

Moon goddess and hunter only at home with wild nature, Artemis returns in women-authored mysteries as varied as those by Agatha Christie, Sara Paretsky, and Lindsey Davis. She even appears in the more domestically oriented modern cozy, for example in her fierce singleness in the quest for justice. According to Ginette Paris, Christine Downing, and James Hillman, Artemis constellates the feminine in her biological incarnation or "the soul meaning of gendered embodiment."[1] She is the feminine of radical autonomy, owing no definition to any context other than undomesticated nature.

Prior to and absolutely outside patriarchal obscurations of woman, Artemis connects willingly only to the non-human. On the other hand, as Downing points out, she is often accompanied by nymphs, and so fosters a protective relation to young women.[2] She is a sister goddess rather than a mother. Artemis protects from her position of virginity, her singleness of being, not from qualities of giving her body in carnal love.

Yet her myth also records that her autonomy, her pure devotion to the hunt, is threatened as is often true of the modern woman in detective roles. In Artemis's myth, Actaeon spies on the goddess bathing naked. His violation is horrifically punished. Turned into a stag, he is torn apart by his own hounds. How many of the determinedly independent women sleuths find their work hindered by challenges to their ability to decide and act for themselves? How often does that thwarting of Artemis *in them* lead to more violence?

Artemis watches over young females and is also guardian of women giving birth, despite, or perhaps because of, not being a mother herself. For Artemis knows more than Western notions of birthing as a relatively safe medically sterilized procedure. Rather, Artemis is the knowledge that the mysteries of birth are also indissolubly linked to those of death.[3] She retains the experience of the long centuries in which childbirth was a major cause of women's mortality.

Giving birth is to be dangerously proximate to the underworld; the presence of Artemis steers the way to death or to life. "The arrows of Artemis... bring death to women," says Downing, noting that the Eleusinian mysteries honoring Persephone are a freeing from the *fear* of death, not from death itself.[4] So many female-authored detectives, and not only bold Private Investigators such as Tess Monaghan and Kinsey Milhone, struggle with their memories of causing or inflicting death.

Above all Artemis is wild. She refuses to make a permanent home with another person even when it means denying her own capacity for lasting love, as V.I. Warshawski does several times in her career. She is wild also in her primary allegiance to nature, as with Nevada Barr's Anna Pigeon. And she is wild in her embodied psychic energy that is kinship with animals and wilderness. Here Hillman reminds us of Artemis as way to the animal within us as vitally (as in vitality), *animated*.[5] We need Artemis' animation because her primitive, skillful embodiment is a defense against forces of the blind, dead power known as Titanism.[6]

Preceding the Olympian cosmos, the Titans were giants, massive, undifferentiated, brutal. Defeating them meant that archetypal principles could begin to stir in the alive, conflicting, mutually implicated, endlessly procreative, and creative goddesses and gods. Personality in all its unpredictable complexity succeeded unfeeling force. Today, suggests Hillman, Artemis energy of animal embodiment is an important defense against any Titanic power threatening to overwhelm the creative polytheism of the psyche.[7] So the fictional detective who takes on, yet cannot wholly defeat, corporate greed and corruption is nevertheless a goddess incarnating vital psychic energies of human being. It is time to look more closely at this autonomous and home-resisting feminine energy.

Artemis Autonomous and Undomesticated

> The best remedy for a bruised heart is not, as so many people seem to think, repose upon a manly bosom. Much more efficacious are honest work, physical activity, and the sudden acquisition of wealth.
>
> —Dorothy L. Sayers, *Have His Carcass*

Upon discovering the murdered body of a young man on a deserted beach in *Have His Carcass* by Dorothy L. Sayers, mystery writer Harriet Vane walks several miles to inform the police.[8] She then calls the major newspapers, so that the subsequent press coverage can double as publicity for her new novel. As she later remarks to her suitor, Lord Peter Wimsey, given that she was herself recently acquitted of a murder, her reputation requires that if she is not to be a victim of press coverage she must herself pursue the case for her own ends. If somewhat forced into an Artemis role of hunting her own hunters, Harriet is, for several subsequent mysteries, more determinedly an Artemis in refusing to compromise her autonomy for the love of Lord Peter.

Refusal of marriage, or the modern equivalent of setting up home with a lover, links a number of female sleuths, including the Hestia-loving Hannah Swensen. Despite having her dream house built for her by sensible suitor, Norman, and despite chasing off his duplicitous ex-fiancé, Dr. Bev, in *Cinnamon Roll Murder* (2012), Hannah resists the temptations of marriage.[9] Not only does she continue to enjoy the attentions of two men (the other being Mike, the more exciting but less reliable cop), she values the autonomy of her own home.

In fact, I suspect there's a connection between Artemis's virginity as being one-unto-herself, autonomy for women, and the unassailable integrity that seems the bedrock of *all* fictional detectives, whether male and female authored. From Sherlock Holmes to Philip Marlowe to Miss Marple to V.I. Warshawski, the fictional sleuth may distain laws, yet always, without fail, pursues justice no matter where it leads.

The fictional detective also never gives up. From Stephanie Plum wrestling a fugitive in the dirt of the streets to the amusement of the local cops, to dogged V.I., who goes undercover in a woman's prison knowing that torture is likely, in *Hard Time* (1999)—the sleuth possesses a dedication to the quest or hunt for truth way beyond domestic or institutional loyalties.[10] Here too the heroism of the fictional detective is embodied; not unlike Artemis in her primary sense of the soul embodied in nature. Moreover, post Sherlock Holmes heroism of the sleuth is characterized by integrity and persistence more than his extraordinary mental powers.

Of course, Holmes does initiate a series of genius detectives that include female authored Hercules Poirot and the devastatingly incisive Miss Marple. Yet Holmes also pursues suspects in the London streets and, in *The Hound of the Baskervilles* (1902), risks being swallowed by a bog.[11] After Holmes, and arguably more emphasized by women writers, detecting talent has been driven less by ego and more by loyalty to justice. Such dedication is enacted in a refusal to relinquish detecting when asked, begged, or even more or less forced to do so.

Fictional detectives by women, from Marcus Didius Falco to Laurie R. King's San Francisco cop, Kate Martinelli, become Artemis in their embodiment of the quest for justice. It is Artemis who possesses Kate with her virginal autonomy in the hunt for killers. They are also Artemis pursuing what has violated virgin feminine nature in the victims of crime. Actaeon invades this pure feminine energy in whatever stops Artemis from being whole, whatever prevents the justice or truth that the Artemis sleuth comes to embody.

V.I. Warshawski with a middle name of Iphigenia appears particularly dedicated to Artemis in her need for independence and constant refusal to compromise. She will not abandon the quest for the truth, even for the sake of those who love her. Losing lover Conrad Rawlings because, even though he's a cop, he cannot tolerate her risk taking, V.I. is constantly in conflict with her father's best friend, Bobby Mallory.[12] He is another cop who wishes she would act more like Hestia.

Kinsey Milhone, formerly a cop, relies so much on her own abilities to read her clients that it seems hard to imagine her not acting wholly upon her own embodied intuition. This particular quality of separateness from institutions in her work is her sense of herself as *whole*, intact in the virginal goddess sense, when pursuing a case. Often starting a hunt with something that cops do not readily do, such as search for a missing person, it is an Artemis-like devotion to autonomy as integrity in the quest that gets Kinsey through.

Indeed, an ambivalent relation to institutions is a fascinating aspect of the feminine sleuth as Artemis. Even though Private Investigators may take on a small job for a large company, as Kinsey sometimes does, most clients are single persons with intolerable problems. Although an exceptional detective like Kate Martinelli may manage to remain with the police, it is indigenous to Artemis detectives to resist corporate or institutional power. In fact being Artemis means hunting alone, even when ostensibly a police officer.

For example, in Laurie R. King's *Night Work* (2000), Kate Martinelli decides to track some clues unofficially and unsanctioned by the supposedly guiding presence of the FBI.[13] V.I. investigates when personally motivated, even though we are told most of her income comes from work for corporations. Typically Sharon McCone enjoys an uneasy relationship with her partner's security multinational that culminates with her detecting its downfall.[14] As James Hillman suggested about Artemis and Titanism (see above), Artemis the sleuth resists, and actively seeks to mitigate, the brute inhuman values of large institutions.

Ultimately, Artemis is the embodied feminine integrity that is marked or violated by injuries in the pursuit. Actaeon appears to the Artemis detective as that which threatens her fulfillment in the quest to know (the pursuit), as well as that which violates the bodily integrity of the victim. Virginal detectives find their own bodily desires and being shaped through their indefatigable determination. So, although Artemis's autonomy is most visible in private investigators such as Sharon McCone, Kinsey Milhone, and Tess Monaghan, she similarly inhabits Maisie Dobbs in 1930s London resisting marriage to her wealthy lover, or Linda Fairstein's Alex Cooper, putting her detecting partnership above the temptation to succumb to love for Mike Chapman.[15]

Remarkably, though, Artemis can survive in some marriages. Arguably Sharon McCone's recent marriage to long-time partner Hy Ripinsky has not compromised her ability to respond to her nature as one primarily questing for justice and helping the vulnerable.[16] Sharon's continuing ability to find her nature alone, in detecting and in wild nature, brings us to other aspects of this fierce goddess.

Wild with Wild Nature

> When Mrs Potter migrated every year from her childhood home in Iowa, then down to the ranch, and then up to Northcutt's harbor, she liked to cook and serve the local fare.
>
> —Nancy Pickard, *The 27 Ingredient Chili Con Carne Murders*

Mrs. Potter, created by Virginia Rich and here continued by Nancy Pickard, does indeed "migrate," for her love of landscape draws her to travel, cook, and care deeply for those who work on her ranch near the Mexican border. When Ricardo, her ranch manager, and his granddaughter, Linda, go missing, Mrs. Potter organizes the search. Their disappearance is then complicated by someone using her famous twenty-seven ingredient chili con carne as a murder weapon.

This sweeping evocative novel emphasizes something present in some other food oriented mysteries, that "home" is not only inhabited via Hestia, however central she may be to works by Joanne Fluke and Diane Mott Davidson. Rather, home also includes a place for Artemis, as a bond to the wild nature surrounding a family dwelling. Mrs. Potter found her ranch with her husband, Lew. Now a widow, she remains embedded in this landscape: "[b]y then there was only one real home for her."[17] Here Artemis and Hestia are allied, even though in this novel Mrs. Potter comes to acknowledge that she has rather too much of Artemis solitariness, and so welcomes male companionship.

In Hannah Swenson's fidelity to the nature of Lake Eden, Minnesota, and Goldy Shultz's love of Aspen Meadow, Colorado, they also echo Artemis. Indeed they both offer a sense of Artemis in their psyche through the detecting process itself. Both women cooking sleuths seem to suffer more than their share of bad or inconvenient weather, as if Artemis is trying to get their attention. Discovering bodies in the snow, as Goldy does in *The Cereal Murders* and Hannah in *Cinnamon Roll Murder*, forces an embodiment in nature that strips it of Hestian comfort.[18] Arguably, Artemis prods some of her more domesticated detectives by combining corpses with weather that demands from them a tougher more embodied style of hunting killers.

Artemis pushes some sleuths into the quest by making first nature and then death get the attention of these busy women. Artemis may also lurk in domestic relationships to animals, showing the capacity to connect to nature other than human. Where V.I. Warshawski acquires two dogs, Peppy and Mitch, through the detecting process, and Marcus Didius Falco a mutt called Nux, Hannah Swensen has a tyrannical cat, Moishe, Annie Darling two tyrannical cats, and Stephanie Plum, a hamster named Rex. Of course Stephanie's hamster is of limited use as a detecting partner; but he does become incorporated into sleuthing when a man threatening Stephanie adds a cat to his cage.[19] Fortunately, Rex's famous soup can proves a suitable refuge from this monster.

In some way Rex stands for Stephanie as an unlikely and vulnerable hero in the chase after bad guys. Given that Stephanie prefers her native smog of Trenton, New Jersey, to anything wild, her sleuthing might appear to range far from Artemis. Yet Rex as animal companion reminds us she is still a hunter, still devoted to her independence from the two men: cop, Joe Morelli, and dangerous Ranger, who adore her. Rex, the hamster, is Stephanie's bond to Artemis as that wild part of herself that refuses to be put into a cage with a wheel, because he is the natural being she isn't. She is not domestic in the sense of adopting the patterns of wife or girlfriend native to her home town.

If Artemis is present in the most urban of fictional detectives, such as Stephanie Plum and V.I. Warshawski, she is surely most visible when a sleuth such as Anna Pigeon, Nevada Barr's Park Ranger detective, takes to the wilderness. Anna works in various National Parks to maintain their safety and the law. In *Track of the Cat* (1993), she discovers the dead body of a colleague who has apparently been killed by feral lions.[20] Anna realizes that she too is potential prey as well as hunter.

> Surely, in this dry season with game so scarce, the lion would return. It might be nearby, waiting. One of the forsworn gods' little jokes: to have Anna's long-coveted first lion sighting be her last sight on this earth.[21]

At this moment Anna is in what might be called an archaic pattern of being equally predator and prey; the conditions probably behind the formation of the trickster archetype itself. Yet she discovers that, like the so-called wildernesses themselves, human culture is inextricably woven into this apparently animal attack. Even when thoroughly tangled in wild nature, murder by human agency can be found by the dedication of an Artemis such as Anna Pigeon.

Here Artemis herself has something to teach; for the deities of Greece do not follow the dualism that severs human from non-human nature. Unlike Judeo-Christian monotheism, in which the divine is creator of nature as separate from himself, Artemis belongs to the human psyche and wild nature as one continuous sacred being. Artemis reveals an ancient continuity between human wildness and the non-human wilderness.

As hunter, Artemis kills and knows why. Her kinship with animals is more than embodiment; it is also survival and, as with Actaeon, a potentially violent defense of her autonomy. Therefore Artemis, like all the goddesses, is not confined in mysteries to the figure of the woman-authored detective. Indeed, her deep connection is present in the *wholeness* of the genre's devotion to the hunt, as the quest to know that which penetrates the mysteries of death and life.

Now it is time to follow Anna Pigeon more closely as she investigates death by wolves, and finds that some savage beasts wear human skin.

Case History (13): *Winter Study* (2008) by Nevada Barr[22]

The novel revolves around a "Winter Study," an actual, non-fictional project to examine wolves on Isle Royale in Lake Superior, ongoing for over fifty years.[23] Anna Pigeon joins a winter study in special (and fictional) circumstances. The new Department of Homeland Security is considering opening the park all year round, a move that would threaten the wolves and end the scientific research. In this exciting and poetic novel, when genuine threats emanate from a man who preys upon vulnerable women, the aim to better protect the Canadian border proves misdirected.

Bob Menechinn, Anna discovers to her horror, drugs, rapes, and photographs women in obscene poses. Anna only finds the damning evidence after Katherine, one of Menchinn's victims, is found dead apparently from wolves, so frustrating the claim that wolves are rarely dangerous to humans. Also sinister are mysterious clues that seem to indicate some monstrous wolf-dog hybrid is stalking the island. In fact two very different "crimes" are colliding to destroy Katherine, who loved wolves with all her being. Bob, her rapist, fails to answer her calls for help because he does not want to go out in the snowstorm. Tragically and simultaneously, Robin, an idealist young researcher, has with her boyfriend Gavin perpetuated a fraud, evoking Conan Doyle's *The Hound of the Baskervilles* (1902). They want to save the wolves and the winter study from Homeland Security.[24]

Faking large paw prints and the huge outline of a wolf seen by Anna from the air, Robin also possesses material irresistible to wolves that she uses in order to manipulate their behavior. Aiming to suggest the existence of a dangerous hybrid that requires the maintenance of the science project, Robin and Gavin mistakenly leave the substance with Katherine, so turning her into prey when she is trapped in the snow. Yet Bob, too, is prey for scientist, Adam, whose wife committed suicide after her rape by the ironically named Homeland Security representative. Anna stops Adam from killing Bob, who repays her by leaving her to die on the ice. Only by setting a trap for Bob, as if he were an animal, can Anna possibly survive this fateful encounter.

Artemis looms large and fierce in this icy series of crimes of predation and hunting. Far more complex than simply Anna hunting to *know*, is Katherine, a violated Artemis only at home with wolves; Robin's purity of desire to save them that leads her into trickily hunting the threatening Titanism of Homeland Security; and Adam's ruthless Artemis-driven hunt of the man who destroyed his wife's psychic integrity.

In the end, the scientific method of hunting truth is not enough. The most dedicated scientist admits that for all the decades of data they do not understand the fascinating, complicated wild nature of wolves. Nor is the scientific method as applied to police detecting and the operation of courts of law enough. While the evidence against Bob is unequivocal, not enough survives his death to convict him. Had he recovered, the material evidence was all too deniable. Anna proposes to the remaining scientists that they falsify their stories and manufacture evidence to protect Robin and Gavin, whom she sees as having "the fearless innocence of young wild things," a wholly Artemis radical defense of the young in the wilderness.[25]

To defend the tricksters who faked the monstrous hound or hybrid wolf, Anna has to get the scientists to agree to abandon their monotheistic adherence to objectivity, valorization of material evidence, and sky father separation between researcher and subject.

As an alternative to objectivity and science, she has to assume an Artemis autonomy from patriarchal systems and set up another relation to nature and human nature, one that will put the wilderness before self-interests.

> "So we play God?" Jonah asked.

> "People always play God," Anna said. "There's nobody else to do it."[26]

In this remarkable novel, Artemis acts in several characters as hunter *and* killer, including in the detective's final trap to kill Bob, or be killed by him. Ultimately Artemis breathes in the scientists, who care so deeply for the objects of their quest, the wolves, even as they try to exclude that love from their research. Artemis is also present in Anna's defense of Robin from Bob's attempt at violation when he drugs her, and in Adam trying to get Artemis's revenge on Actaeon meted out on Bob. Here Artemis is in the rage of and for women raped as well as in the wilderness threatened by the penetration of all year round human pleasure-seekers.

Above all, Artemis is here in the *wildness* of nature. Isle Royale, white, snow-covered and intact, takes on Artemis's fierce virginity in winter with only the careful winter study present to treat it as sacred ground in their terms. Indeed, the unforgiving ice does live out Anna's early Artemis-like thought that "Mother Nature wouldn't go quietly and she would take as many of the enemy with her as she could."[27] Of course when wolves are tricked into killing Katherine, and Robin and Gavin risk prison to defend the pristine winters of the park, Anna finds a more complex inter-relation of human and natural interests.

Artemis is the figure who finally stands for life as natural and nurturing to the child-like and innocent, yet who is also a goddess without regrets at having to kill Bob Menechinn (or be killed), because of his violations of the feminine. The end of the story is Artemis also because saving Robin and Gavin is saving the love they bear for wild nature, which proves stronger than the scientists' so-called objectivity.

> Like a trapped animal, Anna howled.[28]

To save her life and prevent Bob from continuing to rape women Anna has to be both predator and prey. She must think and be embodied as a woman who can hunt as well as be hunted. In Isle Royale and in *Winter Study* itself, female embodiment is explored and tested against wild beasts of human and non-human varieties. Fortunately, the goddess that unites human and inhuman fierceness inhabits this powerfully whole novel.

Sister not Mother Goddess

Artemis is a sister goddess in that she is often accompanied by wood nymphs giving her a protective role with young women. She is not a mother goddess in the sense of having given birth. Perhaps her sustaining quality for women in labor depends on her freedom from such pain, as well as her link to female embodiment and the wild nature we share with animals in birthing. What this means for women's mysteries is profound in Artemis as a powerful hunter who nevertheless respects other's autonomy as well as her own. The Artemis sleuth may lead and initiate but she does not grasp others in possessive mother-love. She is less hierarchical and more cautious of the wild nature in the women and men she cares for.

Such a source of innate strength in freedom for self and other occurs in Jacqueline Winspear's Maisie Dobbs, a young woman in 1930s London. Maisie chooses her own work. She is not to be deterred by the masculine police force, or by the preferences of her sometime lover, Lord James Rowan, son of her benefactors. Her essential solitariness confirmed by the death of a fiancé while a doctor in the World War One trenches, Maisie is Artemis in her self-determination, her use of her own body as a knowing organ, and her sisterly approach to the emotionally and physically wounded she encounters.

Whether tending a badly beaten teenager forced into prostitution, or counseling her bereaved wealthy friend into avoiding alcoholism, Maisie's heightened embodied presence offers a sisterly, non-judging and non-controlling succor.[29] As we see in *Birds of a Feather* (2004), Maisie has learned from her mentor, Maurice, to awaken bodily senses of smell, somatic agitation, and awareness centered by skin and bones to "read" a crime scene or to calm a traumatized witness.[30] From Maurice's Indian guru, she has also learned techniques of embodied meditation that enter the wild psychic powers of the earth in more extraordinary ways.

Even more striking is Maisie's ability to increase skills of empathy, to the point where she can trace deeply buried threads of twisted relationships leading to murder. Why are four post-war women being killed, leaving only a white feather as a signature? Guilty is a woman driven mad by the death of a son who was sent to war by women giving out such feathers. *She* is a dark mother goddess devouring in her rage. Maisie, by contrast, can use a highly developed embodied intuition to detect the unappeasable grief that has burst normal ethical limits in a suffering woman.

Above all, as sister goddess, Maisie as Artemis can embody a connection to wild nature that is potentially healing because it is not limited to conventional notions of family or social codes. For the surviving victim of the avenging mother, Maisie offers restoration of her essential nature. She does so by powerful somatic empathy able to evoke a rootedness in her own body for the troubled young woman. Such a psychic embedding in female bodily being is made possible by Maisie offering a sisterly, more horizontal relationship, rather than a maternal one. By being reconnected to her physical nature as wild, and not *determined* by domestic ties (while still being affected by them), the victim who gave a white feather to her now dead brother may find some peace.[31]

It is a particular quality of Artemis in women's mysteries to provide unique hope for psychic or moral regeneration once profound evil saturates a person or a group. Artemis offers such hope because her crossing of human and wild nature opens possibilities for rebirth in the wilderness beyond human structures. Reborn in animal nature, the Artemis-person is animated enough to try renegotiating the relationships that make us human. It is for this reason that Maisie has been enjoined to close a case with respect for the psychic completion of those involved. Sister goddess, Maisie visits those involved several months later to aid their mending of social and familiar bonds. She is here sisterly Artemis reconnecting witnesses and victims to their need for home in wild psyche that can also enable them to find Hestia.

Of course, Maisie is not alone in embodying Artemis as sister goddess. Private Investigators such as Sharon McCone, Kinsey Milhone, and V.I. Warshawski excel at guarding their own autonomy, in part by respecting that of others. By doing so, the P.I. is an endemic Artemis, in contrast to the police, who inevitably incarnate a paternal codification of masculinity when dealing with victims or suspects. In fact, all three of these pioneering hardboiled detectives exhibit a tension between Artemis as needing to work in solitude, and Artemis as sisterly support to the human and non-human nature in us.

It is V.I. who, in *Toxic Shock* (1988),[32] when enlisted to investigate the unknown paternity of a former neighbor, most overtly tackles this ambivalence head on. With her single mother dying of diseases dubiously connected to a polluting factory, Caroline wants to know if she has any other family. This pull to Hestia however, proves traumatic, as the goddess has been invoked in the underworld darkness of the incest that precipitated the pregnancy. Caroline confesses that she hoped to be revealed as V.I.'s half-sister, child of Tony Warshawski and her own sick mother. Infuriated by the needy child of the past become the difficult client of now, V.I. tries to prevent her own thirst for freedom from domestic ties from driving her to neglect this nymph of the urban jungle.

In the end, shocked by the horrors of the corporate greed behind the mass poisoning of workers, both Caroline and V.I. invoke Artemis as sisters by free choice not enforced by (in this novel, dark) Hestian biological bonds. V.I. hands over the lengthy lawsuit to a lawyer more suited to Athena's care for community justice, while Caroline seeks psychic rebirth in Artemis's woods. She leaves the city to train as a Park Ranger.

Meanwhile Kinsey Milhone's particular ambivalence about the sister goddess is rooted in her female body; that is, her uncanny physical likeness to cousins she never knew she had and who eventually seek her out. Extremely wary of a family who disowned her parents, Kinsey is still moving through novels avoiding an encounter with "Grand," the matriarch who exiled her mother for marrying unsuitably. With a Hestia centeredness in her own tiny apartment, Kinsey is Artemis in her refusal to submit to institutional pressures of police and extended clan.

Lastly Sharon McCone navigates her Artemis ambivalence to her "nymphs" by managing her own detecting agency in later novels. Negotiating close and protective friendships with her employees, Sharon nevertheless detects alone with not even long-term partner and later husband, Hy Ripinsky, eroding her love of solitude and the wild. Hy is something of an Artemis himself, having discovered the inhuman within his psyche in the murkiness of war and drugs in Vietnam. Fortunately, through meeting a woman who enabled him to connect to non-human nature in environmental activism, Hy becomes Artemis-like in an essentially solitary defense of wild nature, as we will see in Case Study 16.

The Defense of Nymphs

An aspect of Artemis as sister goddess is her protective function to young women, her nymphs of the wilderness. Perhaps their incarnation as nymphs signals the pre-marital girl, one more bonded to nature than social conventions. Once initiated into society in marriage, the nymph becomes fully human yet retains her link to Artemis in the biological mysteries of childbirth and death. Here Artemis protects women in their biological, as opposed to social, nature. Here too the sleuth may be such a protector, or she may encounter one whose Artemis passions drive her to fury and murder.

Agatha Christie's Miss Marple, the sleuth as elderly spinster, fulfills many of Artemis' roles, including that of protector of young women. In *Sleeping Murder* (1976), she comes to the aid of Gwenda who is staying in London without her husband, when she begins to fear she is going mad.[33] Taken to a play, Gwenda has an uncontrollable reaction to particular lines that accompany the murder of a young married woman on stage.

> "I was back there - on the stairs, looking down on the hall through the bannisters... She was dead, strangled, and someone was saying those words in that same horrible gloating way - ..."
>
> Miss Marple asked gently: "Who was dead?"
>
> The answer came back quick and mechanical.
>
> "Helen..."[34]

Prior to her hysteria at the theatre, Gwenda has had an uncanny experience while obtaining a house for her new married life in England. Arriving from abroad as a stranger, Gwenda found herself instantaneously attracted to one particular house at first sight. She knew instinctively it was hers, and her Hestia energies were so activated that she was hardly surprised when the interior felt so familiar, so materially animated in her psyche. Yet when she imagined a specifically patterned wallpaper and then found it present underneath a newer décor, she began to experience the house as uncannily alive. Now she is Hestia betrayed. Some terrible secret is binding her to this so-called home, and simultaneously rending her apart within it.

Gwenda fears she is losing her mind; Hestia driven mad. Miss Marple points out that although Gwenda believes herself new to England, suppose, in fact, that the house *had* been her home as a child? Enquiries are made and Gwenda discovers that the words on stage have dug up a real repressed memory, and a real forgotten dead body. Helen was her stepmother, and horrifyingly, Gwenda's father died indicted for her murder. After this discovery, Gwenda is unable to embark upon married life. Can Miss Marple can solve the mystery that appears already to have been decided? To let this sleeping murder lie is to leave innocent Gwenda with the inheritance of a marriage that led to sexual jealousy and killing.

Here Miss Marple as Artemis virgin sleuth is ultimately in defense of female sexuality and protective of the female psyche. For just as Gwenda's initial fear of incipient madness proves an inaccurate interpretation, so does the early impression of Helen, the victim. She is accused early in the novel of promiscuity, yet Miss Marple discovers her to have been loyal to her marriage. In fact, Helen was actually the subject of insane sexual jealousy that led to her death. Seemingly, in retaining her virginity, Helen was Artemis, in the sense of the goddess in preserving her bodily integrity. In this way, her killer was Actaeon, whose desires are illicit and unwanted. He as predator kills and it requires another Artemis, Miss Marple, to avenge one nymph and protect another, Gwenda, from a murderer still at large.

By proving the case to have been unsolved and solving it, Miss Marple has explicitly defended the bodies and psyches of inexperienced young women. In particular, she has validated the ensouled body of women, their sexual being as natural and even virginal in the sense of being whole and entire. In other novels, too, such as *The Body in the Library* (1942)[35] and *A Pocket Full of Rye* (1954)[36], Miss Marple explicitly defends the innocence of immature young women who get caught up in the violent plans of killers. Indeed, in the latter novel Miss Marple is described as an "avenging fury" when investigating the death of a young servant girl previously of her household.[37]

Such vengeful purpose in a sleuth with Artemis's rage at violation of feminine nature brings us to a case study where another goddess, Kali, of fury and destruction, is explicitly invoked in the mystery.

Case History (14): *Night Work* (2000) by Laurie R. King[38]

[I]nside every woman lurks a force of immense power that, when loosed, results in the destruction of men, that longs to trample even the most beloved of males underfoot, to wade in his blood and eat his carcass...

Perhaps that was the point: that even an ordained minister with a pet dog named Mutt, a weekly salary, and a mortgage could feel that urge, primal and terrible.

—Laurie R. King, *Night Work*

The Hindu goddess Kali haunts *Night Work* as the explicit image of female incandescent rage at abuse by men. She is first a bloodthirsty painting found at a women's refuge center that may be connected to the murder of a violent husband. She is also the subject of a doctoral dissertation by Roz, a minister, friend of police officer Kate Martinelli and her lover, Lee. Roz also seems to be linked to this case because of her own anger on behalf of terrorized women.

Complicating the murder are circumstances that resemble the activities of a group calling themselves the "Ladies of Perpetual Disgruntlement." These anonymous activists have been kidnapping known rapists and women beaters and subjecting them to public humiliation, such as having their sins tattooed on the bodies before leaving them tied up naked in a public place. The "Ladies" do not permanently injure or kill; or do they?

Roz Hall acts as chaplain to the victims of violent men at the women's refuge, including Emily Larsen, wife of murder victim, James. Her dissertation argues that some of the pre-monotheistic goddesses of the east, such as Kali, became absorbed into the Judaeo-Christian tradition through the ancient Hebrew scriptures. Psychologically, she forthrightly proclaims the potential of Kali in every woman. Kate has to decide whether her friend's frequent pronouncements on the divine potential of women's fury could prove perilously inciting. Might Roz's wholehearted endorsement of Kali in the psyche actually inspire, in-spirit, the vengeful killing of abusive men?

Moreover, Kate's feelings about the spectrum of attacks on women are sorely tried when Roz gets her involved in the death of a young Indian girl in an arranged marriage. Was this a case of bride burning such as happens in India when dowries are deemed unsatisfactory? When the young husband is found stabbed, this particular murder investigation converges with the killings of the other abusive men. Kate is faced with a possible chaotic world in which friends turn into savage goddesses carrying out almost indiscriminate revenge.

Kali is not Artemis. Archetypally, culturally, and mythologically different, Kali's volcanic rage is Titanic in consuming all in her path. Artemis, by contrast, discriminates. She pursues a specific figure, Actaeon, to his death for his violation of her autonomy. She brings death to women in their biological nature. She is not hatred of another gender, nor is she the Furies, the Greek emissaries of revenge in a broadly applicable sense (see Chapter 5 for the Furies). Indeed, it is indicative that the Furies are plural, and are less differentiated than the goddess, Artemis, who exists in relation to other divine biographies.

While Artemis and Kali can both channel rage at violation, Kali scorches the earth and does not stop to ask questions about the males she meets. She kills them all and wears a gory necklace of their skulls. In a real contrast, Artemis is a skillful hunter in touch with wild nature. This goddess harnesses her hounds to her discriminating task and embodies the virgin, autonomous generativity of the earth; she does not destroy it.

In *Night Work*, Kate's task is to distill skillful Artemis from the terrible possibility of Kalian influence on the deaths before her. In order to succeed, Kate has *to be* Artemis and to *discern her* by separating out events from their apparent similarities. She has to resist her anger and become Artemis in discrimination. She can then see Artemis at work in the crimes ranging from the public humiliation of guilty men to a specific murder, and not in the more Kali-like motivation to kill *all* of them. Artemis is about a knowing and distinguishing of guilt, instead of a Titanic rage at all men for the abuse by a minority.

Kate and her male cop partner, Al Hawkins, are also both driven by the Artemis desire to protect the vulnerable. So they need to be the best hunters. They must be better than the competing FBI so they can track down what appears to be a growth of Kali-like serial killing of men accused, but not convicted of, harming women.

Laxman Mehta, the murdered husband of the burned bride represents the possibility of Kali's horrors realized in indiscriminate slaughter without a hunter's precise skill. Easy to blame in terms of public perception of Indian bride burnings, Laxman was actually a retarded young man who loved his wife. Vigilantes have killed the wrong man. It was his avaricious brother and wife who were bullies. Stealing Laxman's inheritance was their spur to murdering the young wife.

Fortunately, the Ladies of Perpetual Disgruntlement are found *not* to have moved from targeted Artemis-like unconventional punishments to a Kali-indiscriminate killing spree. Restricting themselves to public shaming, they are, however, linked to the fiery Roz. For the "Ladies" are the brainchild of her partner Maj. Kate convinces Roz to tell pregnant Maj to retire them.

Yet there still remain serial killers in *Night Work* who are hunting men merely accused of abuse. Potential victims are located from lists of unconvicted men on a website. Such an approach to revenge is terrifyingly open to mistakes. Kate and Al discover that for these bereaved women, their initial Artemis-like hunting of one specific violator or Actaeon, has gone far further, in unleashing a Titanic Kali. Roz, innocent of wrongdoing in the eyes of the law, may have been the crucial carrier of Kali as symbol of rage for two women who went from killing the man who injured them to systematically executing listed men.

This exciting novel culminates in a fire and a shooting at the home of the immolated Indian bride, with Roz still unable to see that in her passion to save women she is unleashing indiscriminate Kali rather than skillful Artemis. Kate tells her lover, Lee, who is closer to Roz than she is, that Roz is pursuing the bride burner because she herself inspired the vigilantes with Kali.

> "…I think we'll find that she introduced them to the goddess Kali as a feminist avenger and they ran with it…Roz is about to set loose a tornado on the city that'll make it nearly impossible to investigate the Mehta case with any hope of

conviction, and might well drive the Mehtas back to India
and out of our jurisdiction..."³⁹

Driven Roz is still a vehicle for Kali's passion. As she later tells Kate,
she does not consciously condone murder, even though she may have
inspired it. Yet her fury at the burning of Pramilla Mehta assumes with
Titanic indiscrimination that this is simply yet another instance of the
horrible Indian practice of killing brides over dowries. Such an
assumption is unhelpful to this particular case, since the truth of the
young woman's demise is actually far more complicated than it appears.
Pramilla's death is more focused and precise than historical horrors. It
follows from the dysfunctions of one family far more than a lingering
yet banned practice in India. So accusations of bride burning that
would evoke racial prejudice would most likely result in the guilty
escaping the precision of the law.

Kali is no detective. Her rage is promiscuous rather than
discriminatory. Here we see the difference between Kali and Artemis
as that between the Titans and the Olympians: the Titans do not
know; they only inflict. Kate needs to *know*. She needs the truth
that will be more than a collective blind rage. Kate becomes the
virgin hunter Artemis, especially when trying to stop Roz unleashing
Titanic Kali anger, which would actually prevent justice for a vulnerable
nymph, Pramila Mehta.

Artemis is indeed counter to Titanism, as James Hillman has
suggested. Kate and Al stop the Kali serial killers who have
mistakenly murdered at least one (mostly) innocent male. In *Night
Work*, Artemis defeats Kali, and in doing so saves San Francisco from
a storm of racial anger and the fear of vengeful women targeting
men without discrimination.

Turning from the still pertinent issue of men hurting women, it
is time to look at the role of Artemis as the soul of the female body.

The Soul of Gendered Embodiment

Artemis is the psyche as embodied for the feminine. We are in
nature as biological beings who menstruate, bleed, need food, breathe,
experience pain when hurt, and so on. One aspect of the sleuth in/
with Artemis as the soul of gendered embodiment is how the hunter

becomes the hunted. She experiences in her body the knowing of being both pursuer and pursued.

For example, with Linda Fairstein's Alex Cooper there is an almost ritual quality to her regular pursuit by the very killer that she has aggravated by her investigations. Her indefatigable cop partners, Mike Chapman and Mercer, rescue her from being walled alive in *Entombed* (2004), from being murdered in an abandoned plague hospital in *The Deadhouse* (2001), and from death in an abandoned fort in *Killer Heat* (2008).[40]

However, Alex does not passively wait for male assistance. While being unable to physically overcome her attackers in a straight fight, she uses her intelligence and her body to try to escape. In *Killer Heat* she manages to stab sexual predator Troy Rasheed in the chest and warn Mike of the danger he still poses.

> I didn't want him to encounter that wounded animal, still armed
> with Mercer's gun.[41]

Artemis forbids leaving an animal bleeding and in pain. Indeed, wounding by Alex goes against Artemis, for it means that Actaeon is still violating this hunter; still hunting her. To be Artemis she has to kill or at least immobilize. Here Alex as Artemis protects her virginal privacy, her ability to be whole in pursuing her sexual offender cases. Of course this includes protecting her own body, just as she uses her gendered embodiment to reach out to abused women. Turned from hunter and protector of nymphs to being hunted by Actaeon-like predators, Alex is viscerally aware of her body's limitations and capacities. In *Entombed* and about to be buried alive, she has to be Artemis in making a clean kill, or at least effective enough to strike and disable her opponent.

> There would be no second chance for me. If I didn't make a clean
> strike, it would be my own, very premature burial.[42]

Another hunter who is also hunted is Sam Jones in Lauren Henderson's *Chained* (2000).[43] The novel begins with the sleuth waking up chained in a basement. Fortunately for her, she has been left vital sustenance in a flask of instant coffee.

> The fools! Didn't they realize how dangerous it was to give me
> caffeine? Already my headache was fading and the blood was

racing faster. Forget your beaker full of the warm South, this was
the business if you found yourself chained to the ceiling of
someone's back basement.[44]

Chained charts Sam's incarceration while she's remembering the
events leading up to her kidnap. She experiences the indignity of
spending days in dirt and without toilet facilities, followed by threats
of rape from one brutal captor. But Sam has prepared for this last horror,
as she has managed to prize up a flagstone and is waiting to use it. A
long fight ensues, and Sam just manages to get away before being
knocked out while running from the house.

Dumped in a London park, Sam wakes up in hospital to find her
appearance a shock to boyfriend, Hugo, and ex-lover cop, Hawkins.
They tell her that while she was imprisoned a man she was working
with had been horribly murdered; and that it seems connected to the
animal rights threats that led up to her kidnap. Knowing Sam, and to
distract her from her injuries, Hawkins has brought the gruesome crime
scene photos. Paul has been disemboweled.

> It was a constant effort of will to remind myself that what I was
> seeing was a real person, slashed open like an animal in a
> slaughterhouse.[45]

Sam has just spent half a novel in psychic and physical
confinement. Her visceral narrative nevertheless falters in the face
unendurable and unmediated physical suffering: "[h]ard to believe
the human face could distort itself that much."[46] Here the hunter
who is hunted comes up against the victim, whose death the sleuth
may come close to appreciating but not fully experiencing. Sam's
bodily inscription of suffering shocks those who know her and acts
as an embodied metonym (connecting her psyche to the dead
victim) for the unknowable pain of Paul.

Similarly, in Lindsey Davis's *Three Hands in the Fountain* (1996),
the sleuth connects viscerally to the dismembered victim.[47] Someone
has been killing young women and placing their body parts in the
water supply of Rome. When a third hand is found in a fountain,
it is brought to Marcus Didius Falco and his wife, Helena, in order
to engage their help.

> Helena Justina has reached out abruptly and covered the severed
> remains with her own much paler hand, fingers splayed and

straight, thankfully not quite touching the other. It was an involuntary sign of tenderness for the dead girl. Helena's expression held the same absorption as when she made that gesture above her sleeping child.[48]

In mysteries, Artemis as the soul of gendered embodiment becomes a form of knowing. When the hunter is hunted it is a knowing of the killer, his methods, his embodied psyche that extends beyond the villain himself. For in being hunted the detective encounters through her body both murderer and victim. Artemis offers a corporeal knowing that is more than empathy for the victim. It is a metonym of suffering endured by the unknowable death that is dramatized in the novel. The corpse is not just a lump of flesh without animation or humanity. Rather, divine Artemis can animate knowing through her body an intuition of the mysteries surrounding such a death. Artemis, divine hunter, knows in her body what it is to be hunted; this makes her a great detective.

Lastly here, Stephanie Plum begins *Hot Six* (2000) trying to talk a person who failed to appear at a bail bond hearing from jumping from a bridge.[49]

> "Make your mind up about jumping, because I have to tinkle and I need a cup of coffee."

> Truth is, I didn't for a minute think she'd jump…You don't jump off a bridge in a four hundred dollar jacket. It isn't done. The jacket would get ruined… [I]n the Burg you gave the jacket to your sister, then you jumped off the bridge.[50]

Gendered embodiment means knowing how to connect through our shared corporeal life, including how the body is psychically energized by clothes and social customs. Stephanie's rootedness in the New Jersey city extends to intricate gendered practices of adornment that prove surprisingly integral to doing her job as bail bond enforcer. By invoking her body in a way that connects viscerally and socially to distressed Carol, Stephanie talks her down.

Now for an Artemis whose moonscape inspires, in-spirits, magic.

Case History (15): *Freeze My Margarita* (1998) by Lauren Henderson[51]

> For a moment I was in the enchanted forest, a wood
> outside Athens, and then the scent of Sophie's
> cigarette, trails of smoke rising through the branches
> and twining through the sunlight, brought me back
> to the present place and time.
>
> —Lauren Henderson, *Freeze My Margarita*

A chance encounter in an S and M club brings a commission for solitary sculptor Sam Jones. She is to make ethereal hanging mobiles for the set of a London theatre production of Shakespeare's *A Midsummer Night's Dream*. There is a lot of Artemis in Sam, since she fiercely rejects conventions in her sexuality and in her art. She determinedly lives alone, outside patriarchy and monogamy, in squalor that shocks her more Hestian minded friends. In *Freeze My Margarita*, such dedicated autonomy complicates her growing attraction to lead actor Hugo, who plays Oberon, the King of the Fairies.

Pervading the novel are Sam's moon-like sculptures, large metal artefacts evoking Outer Space. Yet Sam's deepening interest in the creative hive of the theatre begins to change her work.

> Before my style had been based on meteors, comets, undiscovered
> planets, wrapped in wire and hung about with strange rings.
> But using the ivy from the cyclorama had started me on a new
> track, aided and abetted by [assistant] Lurch.[52]

Sam's set for *A Midsummer Night's Dream* blends an unknown and inhuman cosmos with a magical forest. Moonlit and brimming with the potential to transform beyond everyday human possibilities, it is a place of the gods; their nature, not ours. Sam's Artemisian powers directed to magical creativity shows her as moon goddess, hunter, and ruthless in her pursuit of her art.

> [I]n the foreground the mobiles floated, the glimmering steel-
> blue of platinum, almost haloed in their own light, seeming
> neither organic nor planetary but some strange new hybrid.
> Trails of silver ivy twisted around and through them as if they

were meteors which had fallen into a magic forest and become
entangled with enchantments.[53]

Here is Artemis outside the human and patriarchal world, a state
also evoked by references to Shakespeare's play containing fairies and
elemental spirits of nature in a warring King and Queen. Unfortunately,
intense conflict is not confined to the play; it erupts in acting rivalries,
personal jealousies, and the discovery of a corpse floating in the flooded
basement of the theatre.

The dead nymph proves to have been an aspiring actress. Suspicions
veer to the theatre's lecherous director, until he is discovered murdered
in his office. The play must go on, and does. But whose ambition and
desires are so out of control as to kill for what theatre stardom can bring?
Who is so determined to become a god of the theatre that they imagine
they can behave as a god with human lives?

Freeze My Margarita looks at the human cost of inhuman and anti-
social desires. Artemis in her fierce singleness, Sam fears the threat to
her virgin wholeness by a real relationship; a tension with Hugo is
maintained in several succeeding novels. However, she is also Artemis
in protective feeling for the vulnerable and those who become her
friends. These are the nymphs she tries to help in investigating malicious
tricks played in the theatre that seem to be escalating into murder.

Yet there is also Artemis in the cruel tricks and killings, in the
inhuman pursuit of theater magic. The goddess's lack of patriarchal
limits can also be a fatal absence of human restraint. Artemis does
not put others first when forced to relate to them, such as in the
collective enterprise of the theatre. From the bitchiness and jealousy
of some actors, to attempts to hurt one or destroy a performance, to
murder, malice stems from an Artemisian distain for human
conventions and feelings.

As bringer of death in hunting, Artemis is arguably present in the
killer who is without normal limits in pursuing his desire for an actress,
moreover one seemingly without human emotions. Alive only in the
fierce moonscape of her art, Hazel, the actress, is supremely talented,
perhaps because of a dedication to work that doesn't afford her a personal
life. She, too, is Artemis. She too is what Sam might become if she
allows her art to drive her beyond human love. For Hazel is hunting
career affirmation without any real concern for, or even understanding,
of the emotions she can arouse as a person. Hazel and Ben, who kills

for her, both live outside human communal consensus. They are wild, living as god(dess)es in ways that bring death to others.

Sam, on the other hand, takes Artemis as the need to be unconfined and feral in her passions. She is Artemis as virgin needing her separate authentic being and also by struggling to maintain this psychic condition of wild animal freedom when finding love. Yet Sam has and needs friends. She is Artemis as protector of nymphs; those younger or more innocent, as in her new assistant Lurch, a theatre apprentice she watches over. Her protective drive ultimately works creatively with her need for autonomy. It pushes her into detecting because she knows that in this theatre world she, an outsider with a past, is a likely suspect. Moon goddess as sculptor sleuth, Sam Jones hunts murderers to preserve her virgin nature and extend a divine protection that tries to conceal her capacity for love.

Artemis: Bringer of Death and Sacrifice

As hunter and therefore killer, Artemis, as *Freeze My Margarita* above demonstrates, is alive in more than the protective drive of the sleuth. And sometimes the detective takes on a darker Artemis in feeling driven to kill, or not to prevent a likely killing. Ginette Paris's discussion of Artemis's archaic legacy in human sacrifice pertains to both detective and prey in mystery fiction.[54] Indeed, one distinction between sleuth and villain may be that the detective considers self-sacrifice, while the murderer usually does not. So what does that mean for the sleuth who considers that the only way to protect the innocent is to kill the predator?

For example, in Margery Allingham's *Death of a Ghost* (1934), Albert Campion comes to realize that a beloved friend, widow of a great artist, is threatened with murder.[55] There seems to be no way to stop the killer short of his execution. Luckily for Campion's conscience, the killer attacks Campion before he makes a move, which leads to the killer's downfall. Less fortunate is Kinsey Milhone, who in *K is for Killer* (1994) gives the name of the murderer to someone who swiftly organizes revenge.[56] Milhone does this because the killer of the initial victim went on to murder a young prostitute who became Kinsey's friend. The double murderer unluckily cannot be touched by the law. Kinsey becomes a deadly Artemis avenging her nymph.

There is, of course, a larger sense in which the sleuth kills by the very act of catching murderers when the penalty of the law is death. Most mysteries do not concern themselves with legal punishment unless convinced of an injustice they are seeking to overturn. An exception is Dorothy L. Sayers, whose Lord Peter Wimsey stories are all set when England still executed convicted killers. Not only does Lord Peter explicitly note that, in the family complications of *Clouds of Witness* (1926), he faces the choice of sending his brother or sister for hanging, he also meets his future wife when, in *Strong Poison* (1930), she is facing a death sentence for murdering her lover.[57]

Although all the people close to Lord Peter conveniently prove innocent, we do witness his agony when the guilty are executed. In the final episode of *Busman's Honeymoon* (1937) and tangentially in other novels, Lord Peter breaks down when "his" murderers are hanged.[58] Sending someone to their death ignites in this sensitive sleuth his traumatic adventures in the trenches in World War One. As an officer, Lord Peter ordered many innocent young men to kill and to be killed. He ends the war too shell-shocked to give orders to anyone. Yet his treatment of others on these killing grounds is indirectly responsible for his own return to health when after the war he is nursed by devoted former soldier, Bunter. This loyal "masculine ex-nymph" becomes Lord Peter's manservant and detecting sidekick.

Interestingly, Lord Peter and the novels themselves maintain an Artemis-like acceptance of the death penalty as necessary to stop murderers, despite the rather high incidence of the innocent being accused. In *Gaudy Night* (1935), Harriet Vane, formerly sentenced to hanging and saved by the successful detecting of Lord Peter, nevertheless defends the legal execution to the skeptical academics at her old college.[59] To one who says that the death penalty is "savage and brutal" she retorts that the killer of the victim she discovered was "a cunning, avaricious brute, and quite ready to go on and do it again, if he hadn't been stopped."[60]

Artemis shows a troubling and trickster-like blurring of boundaries between sleuth and murderer. Artemis kills in her hunt and the fictional detective is often not wholly innocent of causing death in her/his pursuit. Such lethal activities may be a deliberate protective act, or facilitation of the death penalty, or the detective causes harm or death just by involving the innocent with dangerous people while solving the crime. It is time to look at how Artemis enables the detective/hunter to survive such ordeals psychologically in animating through animals.

Artemis Animation with Animals

> This sense of the world as an animated being, as a living animal, is the first component of curing enormity. The animal sense engages the world as highly specific and particular.
>
> —Hillman, *Mythic Figures*

All fictional detectives are Artemis in their ability to be animated, as in to invoke heightened bodily senses and skills, as an animal, in the hunt. For them Artemis, with her connection to animals, is the ability to enliven the world by being animated bodily through clues. Furthermore, in psychically embedding us in a textual landscape of unknown potency, these novels enact Artemis *for the reader*. With the detective we hunt for clues. The quest animates our animal sense of specificity and particularity. Here the structure of the story encourages us to examine textual hints and possibilities.

A mystery novel works by revealing its ingredients as clues, signs that become symbols in the Jungian sense as images of something not yet known or that can never be fully understood. In this sense the mystery is Artemis as a process of ensouling matter, psychically mobilized in the quest for meaning and truth. In the detecting story, a metonym for this process that sleuth and reader mutually undergo often occurs in the role of animals. Animals bring a further level of awareness, making more alive, arousing, in an Artemisian sense, the questing psyche.

So we have textual time devoted to the antics of Hannah Swensen's cat, Moishe as he teaches her loving observation in the gentle wilderness of Lake Eden. Annie Darling, in her rural not-quite-paradise of Broward's rock, has two cats whose personalities teach her about mystery in the sense of a nature unlike her own, and indomitability. In Aspen meadow, indefatigable Goldy Schultz has a cat, Scout, who reminds her of the processes of living outside her troubled world. She also acquires a bloodhound whose fallibility in tracking echoes her own struggles to detect. Nux, an ancient Roman stray dog, earns her place in the Marcus Didius family by rescuing the detective from a brutal beating.[61] Her later struggles to give birth to a single pup are watched over anxiously by the family in ways that echo their own proximity to natural death in childbirth and unnatural death in a corrupt empire.[62]

A final example is Nevada Barr's wilderness-based Artemis sleuth, Anna Pigeon. Dangerously close to alcohol dependence, her connection to an animal proves more nurturing.

> Anna traded the wine bottle for the cat, buried her face in the soft fur of his neck, and breathed deeply. "Ahh. Thanks. I needed that."[63]

In the following case study, Marcia Muller's solitary Artemis sleuth, Sharon McCone, meets future partner and later husband Hy Ripinsky in a wilderness setting. By inflicting death or being the cause of bringing destruction to others, both are wounded by their darker Artemisian pasts.

Case History (16): *Where Echoes Live* (1991) by Marcia Muller[64]

> Tufa Lake… These knobby pinnacles of calcified vegetation… have gradually been revealed as the lake's feeder streams are siphoned off for the faucets and swimming pools of southern California… [T]his is a place of great silence… and when a gull cries and launches itself on a steep trajectory into the sun, the sounds reverberate like gunshots off the surrounding hills.
>
> This is a place out of time – a place where echoes live.
> —Marcia Muller, *Where Echoes Live*

Two perspectives on the wild nature of California dominate the opening of *Where Echoes Live*: it is a wilderness indifferent to human time and concerns, and yet also fragile, vulnerable to human greed. Echoes live here in the history of gold mining made substantial in the abandoned town of Promiseville. Amongst traces of the feverish and unsustainable rush for gold, a few reclusive miners, like the woman known as Tiger Lily, still prospect for what little remains.

Echoes live additionally for a Chinese family, now owners of the huge Transpacific Corporation who seem to have ties to a past of exploitation of Chinese workers. Is this why the company has bought land in mysterious circumstances?

Echoes also persist for those driven to try to protect this beautiful and bleak wilderness, notably Hy Ripinsky, whom Sharon McCone meets for the first time when enlisted by friend and lawyer, Anne-Marie Altman, to investigate on behalf of an environmental organization. Hy, Sharon learns, has a dark past in Southeast Asia, where war and drug smuggling merged into an unholy alliance. The wilderness and a woman named Julie devoted to its protection rescued Hy. Married to Julie, his dangerous exploits became oriented toward saving the eco-system. But now Hy is a widower, and when she overhears a possibly incriminating phone call, Sharon wonders if he is in league with Transpacific.

In fact, the mole in the environmental group is another colleague, Sanderman. He proves guilty of moving a dead body found by Sharon, the discovery of which takes her back to San Francisco to question the corpse's widow. The dead man had been involved in shady dealings that included duping his wife's father into selling land to Transpacific, ostensibly for a new gold mine. Actually the corporation plans to build a huge resort. When ex-miner and recluse, Hopwood, finds out he has been tricked, he beats up his daughter, kidnaps the head of Transpacific, Lionel Ong, and plans to blow them both up in his abandoned gold mine.

Maddened by loneliness and religion, Hopwood has turned deadly. Sharon pursues his trail into the tunnels, rescues Ong, but cannot stop Hopwood from detonating the mine with himself inside. Sharon has solved the murder and uncovered the proposed crime against nature in developing large parts of it for profit. What she has really uncovered, partly through meeting Hy, is something about herself.

> Danger, I now realized, was the thing that brought me fully alive. Conquering it and my own fear was what gave me a reason for going on in the face of an increasing sense of futility. That was the real truth that I kept from George and the others, who would have found it a shameful addiction. And that was what Hy had intuited and accepted.[65]

Sharon has come to understand what hardboiled detectives face: the wasteland of criminal corruption cannot be healed by her alone in the role of grail knight. In fact, continuing on the quest for justice makes her more vulnerable to its corruption within herself. From grail knight she becomes fisher king; one whose darkness is an aspect of the wasteland itself. In *Where Echoes Live* she learns this hard truth by encountering a group of people who have been broken by aspects of the wasteland: "[p]eople whose dreams all died," as she tells Hy.[66] These people include Tiger Lily, Hopwood, Sanderman, who's been soured by divorce and lack of ability to connect to people, and, maybe Hy too.

Sharon, fearing her deadly dark moon side, tells of her own complicity with death. Previously she almost killed two people, and what horrifies her is that she *wanted to kill them* in that moment.

> "Each time I really wanted to do it. I was completely in control. All I felt was this ice - cold rage. I wanted to… act as an executioner…
>
> [P]eople were there, people I care about. They saw the side of me that I try to keep hidden. And it changed things."
>
> "You're an outsider to them now."[67]

Sharon is here Artemis in the goddess's killer role. This goddess's capacity to end a human's life shocks her and distances her from the friends who are her family. Here she is wild with inhuman nature; a goddess terrible and implacable in her wrath.

However, as this chapter has argued, an Olympian, archetypal goddess is not a Titan-like killing machine. Sharon may be addicted to danger, but she is not addicted to murder. Indeed, it was love for her friend who was being threatened that drove her into a killing rage in one of those instances she confesses to Hy. Threaten Hestia enough and she may call upon Artemis for her protective yet deadly potency.

What Sharon does not fully see here, but arguably *Where Echoes Live* does, is that Artemis, while inhuman as bringer of death, does so because intimacy with death is indivisible from a divine role as protector of life. Not only does Sharon detect on behalf of others, but her own capacity to be alive, to be fully animated, is now bound up with her primal being as Artemisian sleuth. Psychologically, the effects of her job have constellated her as Artemis whose addiction to danger is an essential part of the goddess in her.

Put another way, *Where Echoes Live* opens with Sharon in wild nature where the wilderness is both beyond what is human and yet under threat from us. Protecting the wild Tufa Lake through sleuthing for the environmentalists takes Sharon into the non-human aspect of the cosmos. She becomes Artemis as protector-killer *and* acquires that sense of echoes from a more expansive natural psyche living in her.

The wasteland of which Sharon cannot heal herself is human-made. Becoming Artemis makes her terrible in domestic Hestian terms, yet also alive, animated in wilderness terms. She is not limited to, or by, her failure to save the whole human world. Perhaps that is why in this story she does not accuse Hy of the loss of self she sees in others. For Hy also has devoted himself to saving the wilderness by going beyond conventional means. While Sharon cannot put into words the chance that there is something meaningful in her continuing quest for justice, her crucial conversation with Hy ends on possibility:

> "Maybe there's... something."

> ... He put his arm around my shoulders, and I tipped my head back against it. After a while he felt around and located the last beer. We shared it as we drifted in the silent darkness.[68]

Connected, Hy and Sharon are both Artemis with their killing potential, archetypally configured in a goddess who nevertheless is a protector both of the wild and of the vulnerability of wild being in human and non-human form. Floating together on a dark lake, Sharon and Hy embody a shared understanding of the effects of violence that surmount rational knowing. Together and alone, they will continue to know Artemis, and to know *as* her.

Conclusion: Artemis, the Hunter

Artemis drives more than just the fictional sleuth in the hunt for clues. She is a knowing and discriminating skillful tracker who is skilled outside patriarchal definitions of the feminine and dualistic notions of humans as separate from non-human nature. So Artemis can be an archetypal energy to mediate the proximity of the fictional detective to death itself.

In so doing, Artemis in mysteries resists Titanism and its deadening of psyche. Sleuths such as Kinsey Milhone, Tess Monaghan, and Marcus

Didius Falco know their infliction of death is inevitable in the quest for justice for a corrupt world. It is hard for them to endure. An Artemisian animation of psyche outside social norms and connected at a visceral level to wild and embodied nature keeps their discriminating faculties alive.

Moreover, Artemis arguably lives in the fictional sleuth in their over-riding drive on the quest for justice and their distinguishing integrity. The detective does not give up and does not give in. Her/his autonomy as sleuth questing for justice rejects social, familial, or personal pressures to compromise. S/he is foremost a hunter, and a virgin in the refusal to allow anything to corrupt a purity of purpose. Lord Peter Wimsey won't stop even when his attempt to clear his brother implicates his sister in *Clouds of Witness* (1926). When hunting to save the woman he loves from hanging, he does not hesitate when promising evidence in her favor evaporates, in *Strong Poison* (1930).[69]

Stephanie Plum's greatest asset as bounty hunter and sleuth is that she never gives up. Kinsey Milhone frequently continues to serve her clients even after the task for which they hired her is formally over; as in *C is for Corpse* (1987) and *M is for Malice* (1996).[70]

Here the often exhausting quest takes on another Artemis attribute as the soul of gendered embodiment. Marked in body by exhaustion or injuries, the detective participates, at least in part, with the physical suffering of the victims. Helena placing her hand over that of the severed victim's, as if her sleeping child, reveals the Artemis who succors women in childbirth, women close to death, and the goddess who protects her nymphs to the point of hunting down their attackers.

By arranging the killing of Actaeon, Artemis takes a precise and terrible revenge. Her detectives harness her skilful discrimination as well as her careful negotiation of the underworld and deathly capacities of non-human nature. Artemis is wild, and fictional detectives are wild because they are rarely reined in. Little can stop Sharon McCone or V.I. Warshawski in their virginal and unconventional pursuit of truth. Less noticeable perhaps, is the way Hestian sleuths such as Faith Fairchild for Katherine Hall Page[71] and Lucy Stone for Leslie Meier deceive their husbands and evade domestic responsibilities in order to protect the vulnerable beyond their immediate household.[72]

Above all, Artemis binds mystery novel and reader, especially through the detective. In her corporeal suffering, virginal purity of purpose, and wild evasion of human restraints, the fictional sleuth animates the reader and thereby initiates her into the quest. Artemis is *us*. She is readers as hunters after truth. Through animating this goddess within the reading psyche, we learn to track clues and discriminate the kinds of guilt and innocence not so readily perceptible outside the mystery form.

Artemis teaches us about life *and* death, suffering *and* integrity, as well as discrimination in a skillful embodied life. She initiates us into a wilderness that is differently defined from the traditional human/nonhuman binary. For this is a wildness of being unhindered by patriarchy, an undomesticated feminine possible in human relations as well as with nature.

Artemis's virginity challenges conventions of all kinds; in this she is opposed to another goddess, yet one not absent from mystery fiction. Athena, known so often as a father's daughter is a goddess oriented to community values. It is time to look at the domain of the protector goddess of a whole city: Athena, and her intervention with those mythical beings of vengeance, the Furies.[73] For it is Athena who is poised to tame these unlimited powers of violence, furious powers that are surprisingly quiescent in mysteries.

CHAPTER FIVE

ATHENA'S JUSTICE

Introduction: Athena and the Furies

Athena is goddess of the city, of what makes communal life possible, and decisively of the Father's values of order and reason. Therefore, she would at first appear to be less necessary to women's mysteries than a hunter like Artemis. Yet this chapter will argue that the Athena of Aeschylus's play, *The Eumenides*, is as vital to detective fiction as the famous detective-murderer, Oedipus, of

Sophocles' *Oedipus Rex*.[1] The unfortunate Oedipus is the first sleuth who proves also to be, terrifyingly, the very killer he seeks. Where Oedipus opposes the brutality and mystery of the Sphinx, Athena faces forces of chaos and blood in the Furies. They pursue the matricidal killer, Orestes.

> Apollo: Go where heads are severed, eyes gouged out,
> Where justice and bloody slaughter are the same.[2]

The Furies, "the everlasting children of the night," are powers of revenge capable of ravaging the earth.[3] They hunt Orestes for the death of his mother, even though he was urged by Apollo himself to avenge Clytemnestra's bloody execution of his father, Agamemnon. Clytemnestra, too, was killing for revenge; she inflicted bloody retribution for sacrifice of her daughter, Iphigenia, to ensure good winds for the war with Troy. Can the cycle of family murder ever cease?

In *The Eumenides,* Athena performs two momentous and related tasks in the interests of enabling a human community to survive the deliberate killing of a member. In order to prevent a single crime inciting bloody chaos, there will be courts of law to render judgment. Secondly, the Furies, those everlasting powers of rage and pain, must be given a place in the divine Olympian order. Whereas the Titans (see Chapter 4), with their pitiless deadening force, cannot be invited into the archetypal differentiated cosmos, the Furies can become the Kindly Ones, if they are welcomed as strangers who nevertheless find a home in the Olympian-human psyche.

> Athena to the Furies: If you leave for an alien land and alien people,
> You will come to love this land I promise you.[4]

Although the Furies pursuing Orestes for killing his mother might superficially resemble Artemis's terrible revenge upon Actaeon, these more archaic forces of revenge are not, unlike Artemis, a complex prototype of the fictional sleuth. The Furies, significantly plural and undifferentiated, do not take a skillful and discriminating approach to killing, as does Artemis, the wild goddess of nature's endless complexity. As both Apollo and Athena make evident in the play, once aroused, the Furies will turn *all* the ripe earth into a wasteland. In offering them a home, bringing them inside the archetypal cosmos, Athena is skillfully taming the untamable forces of destruction. The Furies will remain themselves—non-Olympians from an older, cruder set of forces—yet will renounce their wasteland-inducing properties.

> Athena: Let me persuade you,
> The lethal spell of your voice, never cast it
> Down on the land and blight its harvest home.[5]

Above all, Athena *persuades*. She blends the terrible necessity that is the Furies with the reason, limitation, and containment that makes a human community possible. As James Hillman argues, Athena who brings the arts of weaving and pottery, the bridle, the yoke, and mathematics is the mind as a container of those passions and rages (the Furies) that might otherwise let loose unlimited destruction.[6] Athena domesticates Necessity, moving it "from otherworld to this world," when she brings the Furies *home*.[7] She unites in her divinity her father Zeus's Olympian powers of mind with the non-Olympian Furies or Necessity, Greek *Ananke*. She combines in her person reason and necessity, qualities the fictional sleuth has to come to terms with, however aroused by the enormity of the crime.[8]

Even Zeus cannot overcome the Furies in their role as Fate or Necessity: he cannot bring the dead back to life. So by making a treaty with the Furies, Athena weaves necessity into the realm of politics and the community.[9] By so doing, Hillman stresses, Athena becomes goddess of *normality*. Coming to terms with necessity is a drive to model the community according to objective norms. These norms are collective, logical, and reasonable, rather than imaginal, because they have to contain necessity. Such norms encode the necessity of death as a fate of all. So when death occurs individually, even unnaturally, one death must not lead to the destruction of the whole community. Seeing everything in terms of the viability of communal life, as Athena does, means that her normalizing is objectivizing—what Hillman calls "the blind spot in Athenian consciousness."[10]

> She is mind as a containing receptacle which normalizes through interior organization.[11]

Moreover, in proclaiming her allegiance to the male, as her father's daughter, Athena, as Ginette Paris points out, is the feminine choosing to work within patriarchy;[12] or, as Christine Downing emphasizes, the rational and ordering qualities traditionally reserved for the male.[13]

While Athena's masculine preferences may feel uninspiring in her world of collective patriarchal institutions, she herself says she cannot privilege the murder of a woman over that of a man.

What makes the Furies initially less susceptible to reason in *The Eumenides* is that to them the murder of blood kin is more heinous than killing a spouse who is biologically unconnected. From the perspective of the legal definition of murder, or of the prohibition against willfully killing *any* human being, such a principle is irrational.

Therefore in the universe of the Furies human existence is a bleak wasteland punctuated by irrational bloody violence. In the Olympian cosmos proposed by Athena, such as in *The Eumenides,* a court of law enables judgment to combine reason and necessity. In the courtroom, the Furies are honored by their pain being heard; they have their place. However, they are not in control of justice as a collective system that makes the continuance of the community possible. Athena's justice preserves cultivated ground from the ravages of fury that would turn it into a wasteland.

Father Zeus ordains law, and Athena's courtroom supplies the ground where rational organization and necessity can engage together in the interests of maintaining the city and cultivating the land. Yet, when she says that accepting the place of Necessity is essential to psychic vitality, I would argue that Athena is not in a simple sense the mouthpiece for patriarchy. "No house can thrive without you," she tells the Furies, as her persuasion converts their raw savagery into the instinctual energy needed to maintain life.[14]

Here, I suggest, Athena is building the container for reconciling necessity and reason that is detective fiction as a genre, a *container.* She cultivates paternal reason by revealing its limits and dependence upon Necessity. Patriarchy must be exercised with regards to what it cannot do, for its inability to control all psychic or divine powers, in order for human communities to flourish. Mysteries are one of Athena's most skillfully woven tapestries or molded vessels.

Athena and Mysteries

While some female authored detectives actively take on the Athena role mandated by her as judge, such as Deborah Knott in the work of Margaret Maron, more indigenous is the sleuth function of combining reason and necessity to contain the Furies.[15] For, the fictional sleuth *takes the place of the Furies* in seeking the murderer. In replacing uncontrolled rage with a discriminating rational consciousness, the sleuth is Athena sifting for clues and seeking objective evidence, as well as negotiating with irrational forces of conviction and hate. Here, generically, Athena is a strong presence forming the deep structures of mysteries.

On the other hand Athena is often the subject of a detective story as well as its structure in the role of the sleuth. For Athena's normalizing qualities can either be used to detect crimes in a chaotic community, or, in institutions where the fantasy of normality has ossified into coercion or corruption, where corrupt normalizing might in fact be the spur *to* crime. "Neither anarchy nor tyranny, my people," Athena exhorts the Athenians, and her role in detective fiction is often to pull the community back to her "mean," or middle way.[16]

Athena detects when a community is too disorganized, or when too disciplined. The Athenian sleuth may here be less successful than in *The Eumenides* in re-forming the state. Failing to complete Athena's mission can lead to the hardboiled detective's despair as the wasteland engendered by the insufficiently integrated Furies continues in a dysfunctional city.

In particular, Athena's close relationship to patriarchy is replicated in complex and diverse ways in the sleuth's relationship with the paternal as the police force. On the one hand, often the private detective does find support and order via the police. Yet sometimes the cops show a disconnection from "necessity" in the presence of the Furies around a violent death, Such a situation requires an Athenian detective to re-negotiate the social contract.

If the cops cannot stem the Furies by restoring a sense of justice in the community, then the sleuth has to tackle the fury released by murder. Athena in the texts *contains*; she weaves reason and necessity by her art of persuasion. Fictional detectives persuade. They may not always get everyone on their side, but they do at least persuade the reader of their integrity and truth, and that norms of justice remain conceivable in this place, this community.

So I am suggesting that Athena, who supplies weaving, pottery, and the art of communal living (by generating objective norms of reason and necessity), also, in good measure, offers the detective genre her knowing as another form for the psychic cultivation. Athena, in mysteries, shows us a way to bury the dead by incorporating the Furies into the archetypal psyche. She enables readers to judge and to work with norms that may or may not be congenial, but do aim to be a rite, a ritual to accept the necessity of death without unleashing Furies of despair.

No household can thrive if it does not honor necessity (the Furies); and perhaps no psyche can thrive without Athena's arts. In seeking justice in order to maintain the integrity of the communal body, mystery fiction is deeply Athenian—as much as it is of Artemis, the wild hunter, and Hestia, the fires of the hearth. It is time to look for this bright goddess in the intricate tales she weaves.

Athena as the Perspective of Communal Life

> I understand you are to paint Sir Henry in the character of Macbeth. May I assure you that with Pauline's child Panty on the premises you will find yourself also furnished with a Bloody Child.
>
> —Ngaio Marsh, *Final Curtain*

As the attitude blending reason and necessity to make a city or a community function, Athena appears in mysteries to discriminate between healthy and dysfunctional social groups. As James Hillman points out, Athena is a virgin *and* a mother, here converging in the mother as institution, as in the *alma mater* of a university, or Mother Church.[17] Unsurprisingly then, Athena may first appear as her own shadow side, the institution as Dark Mother, uncanny or devouring. Not yet containing light and dark polarities Athena erupts in mysteries, just as her totem owl can represent calm wisdom or doom-laden screeching.[18] Or Athena can surface in a sleuth who is facing such a chaotic social group that detecting is necessarily accompanied by weaving a reconciliation. The hapless detective is forced to attempt to normalize the situation.

New Zealand born Ngaio Marsh sets some of her mysteries in the realm of the theatre, where demands of art and pressures of fame distort viable communal norms. In *Final Curtain* (1947), she combines the theatre with the country house mystery by having a theatrical dynasty commission Troy Alleyn, artist wife of Detective Inspector Alleyn, to paint the patriarch for his seventy-fifth birthday.[19] Sir Henry is to be portrayed in his most famous role as Macbeth, while his actor children and grandchildren gather for the party, and he, unwisely, periodically changes his will.

Particularly disruptive is the presence of Miss Sonia Orrincourt, indifferent actress and mistress of Sir Henry who is fifty years his junior. Amongst cruel practical jokes is the defacement of Troy's fine portrait. When Sir Henry announces his engagement to Sonia, his death from rich food (or was it poison?) is not entirely unexpected, at least to the reader. Troy's letter describing this sad event to her astute husband brings Alleyn and his detecting team for a visit.

In *Final Curtain,* Athena is a strong presence, both in Troy and then Alleyn. Here the artist as painter provides the containing vision for a household marked by avarice, and actors vying for the attention and love of Sir Henry, star performer and chief audience in one. The household contains a mischievous grandchild, Panty, who suffers from ring worm, and a collection of old books, including one on poisons that turns up in odd places.

In the end, artist and sleuth come together, as Troy remembers a key detail that helps confirm the murderer. Perhaps symptomatically, Millamant, killing on behalf of her son, is like the Furies in putting a blood relation above all other familial and professional connections. Through Troy's art and Alleyn's containing interviews, Athena wins over an obsession that has continued to kill.

In sending in Alleyn to normalize or dismantle violently distorted communities that have fermented murder, Ngaio Marsh typically favors Athena. Such dysfunctional "cities" may be actual theatres, in *Opening Night* (1951);[20] or an occult sect using hard drugs, in *Spinsters in Jeopardy* (1954);[21] or even a mummer's play in which the beheading game gets too real, in *Off With His Head* (1957).[22]

In all these mysteries, Roderick Alleyn supplies the perspective of Athena in interrogating suspects, as the goddess questions Orestes in *The Eumenides,* where she tells him to recount his whole story of birth and land.[23] Exposing the irrational currents of emotion surrounding the killing, Alleyn gets the whole story, and then weaves reason with necessity. He is his Father's child, standing for the police and the law as infallible and incorruptible, which is not the assertion of all mysteries. For Marsh and Alleyn, finding the murderer is unfailingly to remake communal life possible by removing someone vicious and unreformable.

For Marsh, as for some later authors such as Patricia Cornwell, the killer is possessed by the Furies and must be stopped for the community to survive.[24] So Alleyn, like Cornwell's Kay Scarpetta, is as much judge as sleuth. Her detecting is inhabited by Athena, in her devotion to maintain her city by finding a place for one who incarnates the Furies in the justice system. Ultimately, both Marsh and Cornwell are comfortable with the death penalty for their killers. While Athena's Furies are persuadable by her divine reason, the brutal killers of Marsh and the serial killers faced by Kay Scarpetta most definitely are not.

However not all centers of communal values in detective fiction are focused in the sleuth, or are proved to be benign.

Athena's Institutions

> "You two were in cahoots during the whole of that unfortunate affair over the Wyndham Case."
>
> "Just as well we were," said Lady B. crisply. "We saved the college face in more ways than one."
>
> —Jill Paton Walsh, *A Piece of Justice*

One recurring institution in mysteries is the university, which brings together the high ideals of learning and community with the volatile gathering of competitive scholars and romantic students. Athena of communal values may wish to uphold her university as an *alma mater* whose nurturing qualities outlast current foibles. Or, on the other hand, normative conventions may manifest in a mystery to demonstrate the university as patriarchal in a pernicious sense. Athena of wisdom is here partial to the male coalesced into cruel discrimination against the feminine.

Both of Athena's types of universities can be seen in Cambridge, England, as revealed by Jill Paton Walsh and Michelle Spring, respectively. In mysteries such as *A Piece of Justice* (1995)[25] and *Debts of Dishonour* (2006), Paton Walsh sends in college nurse, Imogen Quy to support students and faculty at the suggestively feminine named St. Agatha's.[26] She is particularly a friend of the Master whose gentle reign bears out his preference for tradition to continue uninterrupted in "the somnolence of centuries."[27]

In *A Piece of Justice,* Imogen finds herself detecting to protect her student lodger, Fran, when Fran is commissioned to write the biography of a dead mathematics genius. Research for this biography proves unusually perilous, two previous writers dying mysteriously. Meanwhile the college itself only requires the identification of malign activity to be restored to its core identity. Imogen is in perfect accord with its stable Athenian enabling of a lively community to thrive. St. Agatha's does not require fundamental reform, which is not the case with Michelle Spring's St. Bartholomew's, in *Nights in White Satin* (1999).[28]

The white satin belongs to Katie Arkwright, a student at Cambridge's less prestigious university, who disappears at one of the main university's glamorous May Balls. From a humble background, Katie supplements her resources by waitressing work at St. Bartholomew's, where she has previously suffered a traumatic assault, discovers Laura Principal, private detective. In "The Echo Room," so named for a mural in which, due to the indifference of Narcissus, the nymph Echo is torn apart by shepherds goaded by Pan, Katie was forcibly stripped in front of forty drunken men. Subsequent to this attack, and to punish Katie for complaining, the ringleader goes to great lengths to convince the college that Katie is actually working as a prostitute. This false accusation leads to another assault, this time by senior faculty member, Dr. Stephen Fox, who is afterward found brutally killed.

St. Bartholomew's, unlike St. Agatha's, prizes the reputation of the college over the suffering of the vulnerable. Indeed, Laura uncovers a long history of condoning male sexual rapacity at the expense of impoverished women. For centuries, The Spinning House was where university police locked up women found on the streets who merely had to be suspected of being prostitutes. Laura also finds out that many years before Stephen Fox raped a young female college servant, leaving her pregnant. The girl was assured that the then student would be punished; he was not. When this adult woman discovers how he has treated Katie, her years of trauma and depression unleash the Furies.

St. Bartholomew's is Athena whose support of the Father has turned from nurture of community to a normalizing of patriarchy as the bias that facilitates cruelty to women unprotected by money. Fortunately,

Laura is capable of being Artemis to Katie's nymph, and she, at least does not suffer the fate of Echo. Echoes of Cambridge's vicious past linger in the willingness of the institution to condone male assaults on women. While for Imogen Quy, St. Agatha's does remedy past misogyny in very belatedly recognizing the genius of a woman who actually made the discovery claimed by the feted male mathematics star, no such change of heart is apparent at St. Bartholomew's.

As *alma mater*, the Cambridge of Paton Walsh and Michelle Spring reveal both possibilities of Athena as mother in an institutional sense. St. Agatha's, with its kindly Master willing to defer to his wife and Imogen, is Athena whose Father is paternal, devoted to fairness and willing to evolve for it. On the other hand, St. Bartholomew's has normalized patriarchy as *rule* of the Father in the sense of willingness to sacrifice women without social power. For Walsh, benign Athena is part of the perspective of the novel as quest for socially structured justice where the benevolent Master presides in the court of the college. Conversely, for Spring, the Furies are *unleashed* by the injustice of patriarchy incarnated in the college.

Yet crucially, when she talks the guilty woman into confessing to the police, Laura Principal does become Athena taming the Furies. By understanding and sympathizing, Athena gives the murderer what *The Eumenides* calls "the rites of Apollo." In other words, she offers reason and necessity to remove the suffering woman from the Furies who have tormented her since she was raped.

It follows that if Athena offers a coercive as well as nurturing sense of communal values, she is not confined to universities. Apart from the significant paternal qualities of the police, Athena is also to be found as Mother Church in mysteries ranging from religion as a vital force in a community, such as Mary Daheim's Catholic Church, in her series of Alpine Mysteries, and the Bed and Breakfast murders,[29] to an Anglican church in England as the location of a crime in the clerical detective fictions of Kate Charles.[30]

For Charles, her detectives, lawyer David Middleton Brown, with his specialist's knowledge of church architecture and his lover, Cathedral Canon's daughter and artist Lucy Kingsley, find themselves caught up with the slayings of unpopular priests. Here theological disputes reveal again tensions within the divine complexity of Athena. For ancient communal values, such as compulsory heterosexuality, need to be reformed in ways that enable Mother Church to find a middle way between anarchy and tyranny, just as Athena urged Athens, when she faced the problem of the Furies let loose by murder.[31]

Case History (17): *Appointed to Die* (1993) by Kate Charles[32]

> Looking back on them later, the next few days had something of the quality of a nightmare about them. At the time, though, at least on the surface, life in the Cathedral Close seemed to go on much as usual.
>
> —Kate Charles, *Appointed to Die*

Stuart Latimer, new Dean of Malbury Cathedral, is ruthless and ambitious to the extent that he destroys the delicate balance of eccentricity (anarchy), compromise, and firmness (tyranny) often considered native to the Church of England. Latimer wants money to fund a lavish new Cathedral Center that will cement his career prospects; he is prepared to destroy the lives of colleagues and Cathedral dependents to get it. He decides to sell off the historic silverware, flog the rare books in the library, take the tea rooms and gift shop away from the endearing yet disorganized incumbents, and raise rents that will make homeless those who have given decades of service.

Moreover, when the Dean unjustly denounces the choir master as a pedophile in public, Ivor Jones commits suicide to the horror of this ravaged community. Athena as Mother Church is devastated and so enlists sleuths with a particular devotion to preserving these specific communal values. These protectors of Athena's church are Canon's daughter, Lucy, and her beloved, a gentle man whose life prior to meeting Lucy was one of love rejected. Living with a harsh actual mother, David Middleton Brown found solace in the church as his Athena mother.

The Cathedral community is more vulnerable to the arrival of politically appointed Latimer because of mistakes made by the subdean, Canon Brydges-ffrench, who had long worked for, and expected to ascend to, the position of Dean. Devoted to maintaining the benign status quo in Malbury, Brydges-ffrench fatally doctored the accounts of a Music Festival that failed to cover its costs. He was aided by the unscrupulous Cathedral architect, Jeremy, who is pursued romantically by Rowena Hunt. She herself revels in a position of power, chairing the "Friends" of the Cathedral, an association for raising money. Rowena is ambitious enough to do a deal with Latimer. She is delighted take over the tea rooms and gift shop, and does not care that devoted employees of the Cathedral will be devastated to lose their jobs.

Jeremy, however, becomes interested in painter, Lucy Kingsley, who first visits the Cathedral with her partner, David, to help with the Music Festival. Returning when the new Dean is ensconced, Lucy and David are shocked by the cruelty of his attempts to control the lives around him. After the suicide of Ivor Jones, the situation appears desperate. *Appointed to Die* opens with Lucy's father, Canon John Kingsley, listening to the sirens of police cars arriving at the Dean's house after a confrontation between tyrant, Latimer, and Canon Brydges-ffrench. The latter was apostle of the malleable tradition of Malbury Cathedral that affectionately accepted the openly gay couple running the gift shop, and tolerated the awful cooking at the tea rooms for the proprietor's long dedicated service.

Yet when the police and ambulance arrive at the Deanery, it is Brydges-ffrench, not the universally hated Latimer, who is a lifeless corpse. The subdean has been poisoned by doctored crème de menthe Turkish Delight to which he was addicted. Inevitably suspicion falls upon the only other person present, the Dean, who is promptly arrested. However, David and Lucy, while cordially loathing the Dean, are nevertheless unsure of his guilt. Unlike the police, as intimates of Mother Church they can investigate in ways in which this particular temple of Athena finds most congenial.

The police are outsiders to a community that survives on compromise and kindness. Kate Charles's clerical mysteries are based on a Church of England that can fulfill Athena's institutional mothering function by adopting a broad acceptance of those who choose to belong as opposed to strict enforcement of "rules" or adherence to exclusions of the past. So the mother Athena Church portrayed in Kate Charles's novels *can be* one with women and gay priests and that welcomes non-traditional families.

Such a progressive expansion of Athena is for Charles menaced by forces of reactionary power-seeking as well as the human fallibility of those in responsible positions. In effect, Athena as nurturing maternal institution is at war with both tyranny, effectively incarnated in Stuart Latimer, and anarchy, in the human flaws and misdemeanors of its members. Canon Brydges-ffrench flirts with anarchy in falsifying his accounts, and this makes him fatally vulnerable to the coercive power of the Dean. So he feels that he can only truly serve the Athena he loves, the Church as inclusive, caring mother, by sacrificing himself.

> "In dying, he was attempting to preserve a way of life that he had loved, in the only way he thought possible. He had reached the point where he had convinced himself that nothing less than the sacrifice of his own life would be enough to save Malbury Cathedral."[33]

Brydges-ffrench commits suicide in order to make it appear that the Dean murdered him. Even if the truth came out, he calculated that the scandal of being arrested would tarnish the Dean's reputation and damp down his incendiary drive for control. To a certain extent the subdean is successful, despite David and Lucy uncovering the truth. Yet it is mild David who is most likely to push the Dean into leaving, by discovering that ancient land rights mean that the town, not the Church, owns the land on which the Dean's pet project, known irreverently as his "edifice complex," now cannot go ahead.[34] Without such a spur to further greatness, not to mention two suicides amongst his subordinates, the loathed Dean will depart.

Athena triumphs in *Appointed to Die,* and not only in the probable restoration of the Cathedral as metonym of an ideal nurturing maternal Church. David and Lucy are explicitly motivated to detect in the cause of protecting the fragile balances that enable the caring atmosphere of the Cathedral to return to what it was prior to Latimer. They want to restore the "second Eden, sealed off from the wickedness of the outside world," rather optimistically imagined by Canon John Kingsley.[35] Finally, in stopping the ostentatious edifice complex, David acts in the interests of both town and Cathedral—another compromise of communities enabling collective life to survive and even thrive.

It is time to look more closely at Mother Athena's adherence to the Father.

Athena for the Father

Athena's relationship with the father in women's mysteries is more complex and multifaceted than might first appear. On the other hand, a tricky or even duplicitous attitude to patriarchy could be expected of a goddess who is pro the paternal forces of reason, yet aware of its deficiencies in enabling communities to continue.

> Marla sighed. "If you had a father and fiancé like that, and you were ambitious, you'd have to be wickedly intelligent."
>
> "And capable of being devious."
>
> "Well, that, too."[36]

Ginette Paris reminds us that Athena sponsors qualities necessary for a woman to work *within* patriarchy, and that these will have to include stratagems and devices.[37] Ophelia, in Diane Mott Davidson's *The Whole Enchilada*, has a father who is actively preventing her from inheriting the fortune left by her mother in trust for her until her twenty-first birthday. Without a college degree Ophelia does not get the money, so her father refuses to pay for tuition. Instead he pays a worthless young man to court her, in the hope of a quick marriage and pregnancy. Once wed and a mother, Ophelia's trust forces her to support husband and child, giving her father more opportunities to continue siphoning off her money.

Discovering the plot, Ophelia uses her clothes allowance to pay college fees, completes a degree in record time, and hires a lawyer to get her access to the thirty million dollars left by her mother. All is revealed at her birthday party, catered by Goldy Schultz, where the daughter's stratagem triumphs over the father's financial fraud. By working within the system through hiring an astute male lawyer, Ophelia wins the freedom and autonomy possible through a huge inheritance. Ophelia's mother's name was Athena.

Of course not all patriarchal control is so blatantly self-serving and corrupt. Where Ophelia needs Athena qualities to manipulate paternal greed, Sharon McCone, in *Wolf in the Shadows* (1993), comes up against a far more subtle and appealing version of Athena in her employers, All Souls Legal Co-operative.[38] All Souls has been successful in favoring the poor and disadvantaged. It has been the embodiment of Athena as institutional mother helping the underprivileged from within a patriarchal legal system.

Now All Souls is to expand, become more corporate, and to confine their idiosyncratic lead investigator to a desk job. Sharon, far more Artemis than Athena, instinctively resists. Then she is approached by

Gloria Escobar, who has a strategy to use against Sharon's desire for autonomy, all the more powerful for being out in the open. Gloria describes her own impoverished childhood with a mother who entered the United States illegally and suffered in order to give birth to Gloria as an American citizen. She says that Sharon should be prepared to make sacrifices too, like she has had to.[39] Sharon promises to consider the promotion, which is an attempt to make her more like compromising Athena.

Later in the novel Sharon comes to respect Gloria further by experiencing some of her mother's terror in the arena of illegal border crossings. She also informs her brother of their mother's astute observation that she, Sharon, is essentially an Artemis who will not tolerate an institutional Athena position.

> "Oh…[she said] that there's a side of me that's kind of… wild, is how she put it, that isn't going to fit into any of the convenient little niches that society uses to confine people."[40]

Of course even Sharon cannot ignore the institutional presence of the Father, in the form of the police, as well as the lawyers who employ her at All Souls. Indeed, the variety of attitudes toward the police in women's mysteries can be seen as a range of resistances, accommodations, and even celebrations of patriarchy. Sharon McCone, like Kinsey Milhone and V.I. Warshawski, is capable of negotiating with the police via the sexuality of Aphrodite. Like Kinsey and V.I., Sharon has had erotic relationships with cops (see Chapter 6). Perhaps significantly for these sleuths, all finally too independent for institutions, these affairs do not last long, and are punctured by tensions over the P.Is' independent attitudes to justice and detecting.

By contrast, sleuths as diverse as Linda Fairstein's Alex Cooper and Jacqueline Winspear's Maisie Dobbs work harmoniously with forces of law and order; or, at least in Maisie's case, when they get to know her. Even more firmly allied with the police as a dimension of patriarchy are those fictional detectives that *are* the police, such as Ngaio Marsh's Inspector Roderick Alleyn and Ruth Rendell's Inspector Wexford, along with Laurie R. King's San Francisco cop, Kate Martinelli. Yet as we have seen in previous chapters, even if the sleuth is a cop, the novels are capable of an Athena-like strategic mitigation of patriarchy from within. Examples include Kate's defense of the women abused in *Night*

Work, and the dedication of cops in Fairstein's mysteries to support sex crimes prosecutor, Alexandra Cooper.[41]

Suggestively, even cozy sleuth Goldy Schultz records a change of attitude over the patriarchal institution of the police after she marries a trustworthy cop. Prior to meeting Tom Schulz, the police were the forces of law and order who failed repeatedly to enforce the law against her violent first husband, and were incapable of supporting order in her life. Then, as a cop's wife, when faced with Furies of irrational, unpredictable violence, she finds herself in the Athenian position of seeing the limitations of the law driven by reason.

> Perhaps my misgivings about the sheriff's department had developed from the fact that when I was deeply bruised and even more deeply depressed, the cops had been unwilling or unable to lock up the Jerk and toss the key to his cell over the Continental Divide... Marrying Tom and going through the harrowing experience of having him kidnapped by a would-be killer, I'd also come to realize how dangerous his work with the department could be, and how steadfastly most cops carried out their responsibilities. So my attitude had done a complete turnaround.[42]

Goldy changes her mind about the police in an Athena-like testament to the persuasive powers of reason. Of course some women-authored mysteries are fundamentally in favor of the existing social order. For these conservatives, only moderate reform is considered desirable; or they take the position that the elimination of corrupt practices will be sufficient to restore Eden. One such author is Dorothy L. Sayers, who captures the erotic appeal of paternal tradition in the delightfully attractive form of Lord Peter Wimsey. Jill Paton Walsh's continuation of Sayers' character faithfully follows this adherence to compromise between twentieth-century social change and centuries of patriarchal nobility by having Lord Peter succeed his brother as a reforming Duke of Denver in *The Attenbury Emeralds*.[43]

Throughout the work of both authors the police are entirely incorruptible and wholly accommodating to the aristocratic amateur detective. A kindly police force, lacking only in genius, is represented by Lord Peter's best friend, the highly promoted and unimpeachable Charles Parker. In *Clouds of Witness* (1926),[44] their unbreachable alliance is cemented by Parker marrying Lord Peter's wayward sister, Lady Mary, after her unfortunate experience as a murder suspect.

Given the diversity of attitudes to the patriarchal father in women's mysteries, it is time to turn to their function as *container* and device for blending or weaving such socially disruptive conflicts as the injunction to maintain order in the face of unexplained killing. Remember, it is Athena who in *The Eumenides* insists on the persuasion of reason by asking for a *full account* of the crime and its antecedents. Athena is the mystery form taming the Furies of irrational violence by making a container where the truth can be heard, where reason and necessity meet.

Case History (18): *Gaudy Night* (1935) by Dorothy L. Sayers[45]

> "Then you think we can solve the problem by straight detection, without calling in a mental specialist?"
>
> "I think it can be solved by a little straight and unprejudiced reasoning."
> —Dorothy L. Sayers, *Gaudy Night*

Harriet Vane is invited to return to her Oxford Women's College for a reunion known as a "gaudy." There she discovers a vicious anonymous note directed at academic women. Subsequently, she is asked to return in order to advise on an outbreak of obscene graffiti, poison pen letters, and pranks designed to harm the reputation of the college as a whole. Circumstances indicate that the culprit must be one of only a few students, or a college servant, or one of the all-female faculty. Harriet finds that her Athena qualities are summoned to aid this community in crisis.

> There did come moments when all personal feelings had to be set aside in the interests of public service; and this looked like being one of them.[46]

Reason and persuasion, addressed to the amelioration of necessity and the Furies, suffuse the narrative of *Gaudy Night*. Both Harriet and her frequently rejected suitor, Lord Peter Wimsey, are engaged in trying public service. Harriet aids her old college, partly because of the friendship of former teachers when she herself was falsely accused of murder, while Lord Peter is sent abroad by the Foreign Office to try to resolve a European diplomatic impasse.

After one vulnerable student tries to commit suicide because of a campaign of anonymous abuse, a horrified Harriet calls on Peter for help. It is he who urges Harriet to adopt Athena's reason in approaching the case. She is consumed by fears that a community of largely celibate women harbors a diseased mind due to sexual frustration. It is Peter, the man whose marriage proposals she has rejected for several years, who points out that she is confusing her personal difficulties with the malign details of the case. Reason is far more efficacious when used to separate Harriet's ambivalence over passion from the poison pen's vicious hatred of academic women.

> "All these women are beginning to look abnormal to you because you don't know which one to suspect, but actually even you don't suspect more than one."
>
> "No: but I am beginning to feel that almost any one of them might be capable of it."
>
> "That, I fancy, is where your fears are distorting your judgment. If every frustrated person is heading for the asylum I know at least one danger to Society who ought to be shut up."[47]

In *Gaudy Night* both Harriet and Peter take on Athena's role in mediating the Furies that beset this college community. By contrast, the criminal exposes Athena's potential for darkness and doom. At the start, Harriet provides this goddess's prioritizing of the communal. Yet, in being marked by sexual desire, she is also Aphrodite, and so at the college Harriet suffers from the normalizing gaze of her student peers who have led more conventional lives.

The poison pen episodes are at first a distortion of Athena's normalizing, in suggesting that celibate learned women are abnormal and secretly sex crazed. In fact, once the campaign threatens to induce suicide, the Athenian normalizing gaze has taken on its deathly and haunted shade. The culprit is now referred to as the college "Spook," and her activities gravitate toward the final attempted murder of Harriet. Lord Peter finally discovers that the criminal is a woman whose hatred of unconventional academic women has twisted Athena's normalizing into violent rage.

Lord Peter supplies most of the reason and persuasive qualities of Athena in investigating the troubled community of women scholars. In two major scenes he, aided by Harriet, enacts Athena's role, as in *The Eumenides*, by rationally debating and persuading the scared faculty of the fundamental viability of the college as a caring community. After dinner, he embarks upon a discussion of his vocation of detecting in a time in which "his" murderers suffer the death penalty. Some of the scholars are opposed to legal execution on ethical, medical, or scientific grounds.

Later, when Miss De Vine reveals that she had to deprive a male scholar of a job because she discovered he falsified something in his thesis, the discussion of what to do for the best interests of the community versus the individual becomes more specific. Even though the man had a family to support, she will not compromise the scholarly integrity of her scrupulous world of academic research. For Miss De Vine the core of her being is dedication to its Athenian communal and rational values of integrity in research: these are *divine*. Unfortunately, the man whose thesis she rejected committed suicide, leaving his impoverished widow to take up a position as a servant in an Oxford college. Annie Robinson, unmasked as the vicious spook, is Athena's normalizing now consumed by the utterly destructive Furies.

> "You can't agree about anything except hating decent women and their men. I wish I'd torn the throats out of the lot of you…"[48]

Gaudy Night shows the depth and range of this persuasive goddess in the necessity of reason and the reason of necessity. Annie's husband cannot be brought back from the dead, and her inability to accept such "necessity" *rationally* drives her from an outmoded sense of feminine normality (women as wives and mothers first of all) to implacable Fury directed to all academic unmarried women. Annie's inability to maintain hold of Athenian qualities of reason and its role in prioritizing communal values makes her subject to being devoured by the Furies. *Furious*, she lashes out at the college and its vulnerable young students. Rage against the actual woman who injured her husband is not enough. She becomes the Furies in their indiscriminate malignity. So *un*persuaded, there can be no place for her in the college community. Unlike the persuadable Furies of *The Eumenides,* Annie is banished to prison.

When, in the latter part of the novel, he convinces the traumatized women of the college that they do in fact possess the integrity of a viable community, Lord Peter, primarily, carries out Athena's role. Together Harriet and Peter save the educational institution by rational persuasion and by persuading them of their rationality.

> "...I established for a certainty, what I was sure in my own mind from the start, that there was not a woman in this Common Room, married or single, who would be ready to place personal loyalties above professional honor. That was a point which it seemed necessary to make clear, not so much to me, as to yourselves."[49]

Here the detective is strongly and positively Athena, in demonstrating reasonably and persuasively the necessities of a viable community in which collective values such as academic integrity must overrule personal desire where there is a conflict. Not all mysteries so starkly advocate for community survival over individual preferences. Yet *Gaudy Night* shows that communal interests and conventional norms, or traditional values, are not always identical. This novel is strongly on the side of what in its inception was still unconventional: the place of women as equals in academic learning.

Athena as a divine *feminine* here sacralizes community in a feminist manner. It is a feminism working within patriarchy, mediated by Lord Peter Wimsey as profeminist metonym of a privileged and unequal society. Yet the novel astutely explores Athena's limitations and divinity in tackling the Furies that beset the modern psyche.

Athena as Containment

In the fictional small English town of Kingsmarkham, Ruth Rendell provides two worthy examples of a usually admirable police force. Inspector Reg Wexford is a liberal man who upholds the law sometimes in spite of its impartiality, for he sympathizes with the poor and marginalized criminals who have fewer choices in society.[50] His deputy, Mike Burden, is an old school conservative who regrets many of the social changes of the later twentieth century. Together Burden and Wexford debate and explore social tensions while *containing* them in the institution of law enforcement. In their hands, the law is paternal, allied with Athena in her divine function of protecting the city.

For example, in *Road Rage* (1997), Wexford's tacit support is with the anti-road building protestors, while his institutional loyalties are firmly employed in policing the encounter.[51] When his wife, Dora, a protestor, is kidnapped, it is Wexford's capacious and politically imaginative understanding of environmentalists that leads the investigation to discover that the so-called green criminal gang is, in fact, a cover for something different. Burden, by contrast, regards loud protests as potentially subversive and a prelude to more serious offenses.

Simisola (1995), on the other hand, exposes the limitations in Wexford's progressive understanding, as when he fails to take normal precautions in telling the African parents of a missing daughter that she has been found murdered.[52] When the parents turn up at the morgue they are shocked to see the body of a battered teenager who is nothing like their daughter in age and build. Given the small black British community in Kingsmarkham, Wexford's unwarranted assumptions amount to racism previously hidden from himself. The murdered teenager turns out to be an abused slave hidden from sight in a rich house, while the missing daughter is alive and well and merely dodging her family.

Above all, Wexford and Burden enact a containing function that is twofold. Within a mystery, the narrative and its relationship with the law rehearse Athena's role in *The Eumenides* in stopping a cycle of violence. The truth is uncovered; even if all the malfeasance cannot be solved, at least a potential cycle of revenge is averted. Moreover, as another dimension to Athena's containment that makes the city or community possible, the reader experiences—albeit in fictional form—the *containing* of the threat to social existence that is killing without a solution, without *knowing* as Athena demands to know, the full circumstances.

Of course, as James Hillman argues, Athena's containment is also psychological. She contains secrets, family fractures, truths found out but too perilous to community peace to be widely broadcast. Originally the goddess provided pottery and weaving to make the *container* that might hold necessity and reason together. So in Carolyn Hart's opening novel of the Henrie O. series, *Dead Man's Island* (1993), the middle-aged female reporter is called upon to sleuth for an ex-lover who has been sent poisoned candy.[53] Detecting many reasons for wanting to eliminate the arrogant media tycoon, Henrie O. solves the subsequent murder in a way that restores order, but decides to keep some secrets in the interests of the peace of mind of the survivors. Sometimes a community or a family is made possible by Athena's knowing silence.

Indeed, Athena contains even more explicitly within the psyche, as in Judge Deborah Knott for Margaret Maron. Lest a judge might be considered too starchy a sleuth, Deborah not only has her bootlegger father and huge willful family, she also has warring inner voices.

> *"Presents are for Christmas morning,"* decreed the starchy conformist preacher who lives in the back of my head.
>
> *"But Dwight will enjoy playing with it more before Christmas than after,"* argued the rule-breaking pragmatist their housekeeping duties.[54]

Just as Judge Deborah aims to temper the rules of the law with permitted mercy, so mysteries *contain* by weaving together reason and necessity, impartiality and passion, so that society might continue with order and compassion.

It is time to look at a particular example of a woman trying to be Athena while still being rejected by the patriarchal institution of privilege that is Harvard University.

Case History (19): *A Death in the Faculty* (1988) by Amanda Cross[55]

> People believe what it is convenient for them to believe. For me, however, Kate thought, it is convenient to believe in the police. "I saw too many movies when I was young in which good triumphed," she said. "I know that. But since I haven't separated myself wholly from institutions, I have to believe to some extent in their power to operate fairly."
>
> —Amanda Cross, *A Death in the Faculty*

In *A Death in the Faculty*, Kate Fansler has an Athena job to do without being entirely comfortable in the role. As a feminist and a Professor of Literature in a prestigious New York University, she is able to work for women's advancement from within a patriarchal institution. It helps that her life is assisted by her privileged and wealthy family. So when Professor Janet Mandelbaum of Harvard University is found drunk and unconscious in a bathroom at a faculty party, it is not inconceivable that she should ask for Kate's help. What is unusual is the manner of the request; a radical feminist who despises women who have anything to do with men visits Kate on behalf of Janet, a pro-patriarchal woman who rejects feminism.

Someone is trying to discredit the unhappy Janet, who was chosen by Harvard as a "safe" non-feminist woman after an endowment forces them to actually hire a female for the first time. Kate is persuaded to come to Harvard as a visiting scholar, where she finds more congenial company in Janet's ex-husband, Moon, who was also once her lover. Ignored by the male tenured faculty, Janet refuses opportunities to find support in the women students who want her to fight for more equal treatment. After a faculty meeting goes badly, she is found dead of poison in the men's bathroom. Kate is summoned by Clarkville, the misogynist head of department, to view her body.

With Moon swiftly arrested for murder, Kate helps her friend with a lawyer, and begins to question junior faculty and students. Apart from the convenient scapegoat of Moon, the Harvard establishment prefers to believe that radical feminists must be responsible. Particularly vulnerable to an anti-feminist establishment is divorcing mother, Luellen, who was tricked into visiting Janet when she was passed out. Kate quickly discovers that the cruel prank of giving Janet a drink doctored with 100% proof vodka, then calling the feminist commune, was carried out by Howard Falkland, a junior faculty member keen to do anything for advancement and Clarkville's approval. But who actually murdered Janet?

Meeting Janet's brother begins to give Kate a clue, since Bob proves a bore, a misogynist, and a bully. But true to her calling as a literature teacher, Kate finds the answer in scrutinizing two poems by George Herbert that Janet was reading before she died. Rejected by Harvard, Janet was unable to fight the prejudice displayed by seeing it as the outcome of important gendered structures of power. She has lived her life ignoring her femaleness, trying to strip the feminine from Athena. She wanted to be a scholar without a body and a gender, an Athena with no acknowledgement of the dark. Showing the psychological disaster of preferring one god and repressing all others, such as Artemis or Aphrodite, Janet killed herself. Her body was moved after death by Clarkville to try to make things look less bad for his department.

In *A Death in the Faculty* the mystery itself challenges Athena for her adherence to the Father's values, yet ultimately provides reconciliation. Janet Mandelbaum is one aspect of Athena, in her willingness to serve patriarchy if it will in turn honor her as Father Zeus's daughter, a woman without a mother. For Janet fatally expected Harvard to welcome her as a renowned scholar, which she was, and ignore her sex. Instead they ignored her Athenian achievements in the world of reason *because* of her sex. Harvard, a bastion of class and patriarchal privilege, saw itself as a Zeus not in need of his female children as goddesses. They are only permitted as untenured teachers and fee paying students.

Janet is not able to incarnate Athena to the extent of making Zeus, ably represented, at least in his own opinion, by Clarkville, recognize her divine powers. She, in turn, is not able to unite reason and necessity, until she does so by her own death. As we learn in *The Eumenides,* even Father Zeus cannot bring the dead back to life. Janet was unwilling to force a "court" at Harvard where women can be listened to, like Athena hearing the full story of Orestes. She refuses her divine mission as Father's daughter. Yet as Kate begins to envision her death, it is here that Janet enters the sacred.

> "What eventually occurred to me was that it could be read as an invitation to death, that one was ready to join Christ in heaven..."[56]

> "It has to be a gesture towards a man, perhaps even revengeful..."[57]

Only in suicide is Janet ready to oppose the Father; she becomes fully Athena when exposing the Harvard English Department to necessity. Her death means that their treatment of her *as a woman* cannot be wholly dismissed or forgotten. In the end, of course, Kate and Janet constellate Athena between them, as it is Kate who forces Clarkville to realize why Janet died, and that his world of the Father alone is no longer tenable. Janet will have a successor who will be tougher and able to call the Father to account by combining reason and necessity.

Kate, the sleuth, is liminal Athena at Harvard. She is liminal because, not part of this patriarchal order, she can point out how divorced it has become from the communal values of the 1980s which demand that women be heard. A new "court" must be assembled if the Furies of generations of marginalized women are ever to be tamed. Kate is repulsed by Luellen's bitterness against all men after she has been badly treated by her ex-husband. Luellen is the Furies, in that her anger is not directed in a discriminating way against those who have actually injured her. Rather, she expresses a rage against men as a gender that is *furious*, unappeasable, an incitement to chaos.

So it is Kate, very aware that she is Athena choosing to operate from a privileged position she has not earned, who finds a place in the city for these particular Furies by clearing suspicion of murder from Luellen and assisting in her custody battle for her children. Luellen and the feminist collective will continue in Cambridge, MA, as resident aliens in the divine patriarchal orbit of Harvard. So archetypally Kate and Janet constellate Athena in forcing Harvard to the necessity of listening to the women it has routinely marginalized for centuries. This Athena makes a home for the Furies as radical feminists, while Kate lectures on new paths for women's art as "the new forms possible to women in making fictions of female destiny."[58]

Athena, goddess of weaving and pottery, of the integration of chaos that enables "new forms," here offers in the mystery genre itself the necessity of the feminine.

Athena as Normalizing Perspective

> As I walked along Chicago Avenue, a couple coming
> out of a bar tried to offer me a dollar for a cup of coffee.
> Their gift made me realize what a bizarre vision I must
> present – in my running shoes and bedraggled evening
> gown I was an avatar for homelessness.
>
> —Sara Paretsky, *Breakdown*

One function of the mystery genre adopts Athena's perspective by solving what can never be normal if a community is to survive: the act of murder. By discovering the story of the crime and reconciling it with the communal law, detective fiction enacts a normalizing perspective. Of course, what is also apparent is that "normal" is a cultural construction. Mysteries demonstrate that Athena's vision may be dedicated to collective values, but it cannot disguise tensions and conflicts from within those norms, just as Athena in *The Eumenides* struggles to placate the Furies.

For example, the large division in the genre between hardboiled and cozy falls within Athena's territory: can the society be fully redeemed by solving the murder, or is it revealed as fundamentally flawed? Cozies are set in a "good place," ruptured by the crime that cannot be ignored by the gods. In uncovering the true story of the crime, detective and genre both enact Athena in a context of healing that society. The "good place," or Eden, is restored by an Athenian perspective that provides a return to the sacred as collective. Athena re-collects her city, even if it is a rural village.

On the other hand, the hardboiled detective novel provides a more cynical vision of normality, such as Tess Monaghan's enforced acceptance of endemic corruption in Baltimore, and V.I. Warshawski's near despairing reflection on the ideals she once shared with her reporter friend, Murray Ryerson.

> So much time had passed since he and I worked on our first story together, corruption in the Knifegrinders union. We not only hadn't cleaned up the city, we hadn't even made a dent. Instead, fraud had spread along every corridor of American life and had infected the newsroom.[59]

Here the norm is corruption, particularly in the submission to power. It is a vision of the detective and the novel of a return to business as usual, *except* in the crime of murder. As V.I. repeatedly tells the rich and powerful, in *Breakdown*, she will not compromise when it is a matter of pursuing a killer. For the fictional detective, Athena's ethos means that murder can never be a normal part of community life; it is unacceptable. Perhaps this is why the unimpeachable integrity of the detective is so ingrained in the genre. In refusing to let murder go unsolved, the detective always acts on behalf of the community; and Athena is always present in that role.

Even those instances in which the detective allows a killer to escape the penalty of the law is, I suggest, Athenian, in the sense that the detective, struggling with the decision to withhold evidence or the truth from the legal authorities, is acting on behalf of a sense of justice that is more than merely focused on the individual. The sleuth does not let a murderer go simply out of sympathy with his or her extreme circumstances. Rather, like Orestes in *The Eumenides*, the killer goes free because the health of the community would either benefit by the escape, or be injured by the law being indiscriminately applied. Athena is a communal yet discriminating gaze. She is a strategist who can employ the trickster qualities of the genre to offer a normality that rules out murder.

An example of Athena letting murderers go unpunished, but not unjudged, would be Agatha Christie's famous *Murder on the Orient Express* (1934).[60] Hercules Poirot discovers that the victim, stabbed twelve times, was a notorious and unpunished killer himself. The disparate suspects, he suggests, are *all* guilty, having constituted themselves as judge and jury to avenge a kidnap and murder that destroyed an entire household. As an alternative, he offers the assembled suspects and authorities another possibility of an unnamed and essentially untrackable lone assassin.

Once the full circumstances behind the crime are brought into the open by Poirot acting as Athena, the Furies that engendered the execution on the train are stilled. All agree to accept Poirot's strategy; the trickster version of the crime will be offered to the legal system. Hence this collective Orestes, the twelve murderers, will be able to return to their normal lives unplagued by the Furies that drove them to construct an elaborate murder on the train.

Murder on the Orient Express is a particularly lucid example of Athena's normalizing vision being more nuanced than James Hillman allows, when he says that her vision relies upon collectively derived norms rather than imaginal ones.[61] In fact, in this instance Poirot attends to the Furies driving the murderers, an imaginal invasion, rather than to the legal system's inability to imagine justice in these terms. Similarly, a detective like Jacqueline Winspear's Maisie Dobbs also attends to collectivity on an imaginal level, by trying to weave a new wholeness of bodies and psyches for survivors of a case. Once the investigation is ostensibly over, she visits its scenes again to try to make the threads exposed by sleuthing connect into a psychic container for her clients.

In the person of the detective, Athena's normalizing perspective encompasses psychic as well as social integrity, including I suggest, on the imaginal level. A final example of this occurs starkly in Ruth Rendell's *The Veiled One* (1988), in which the collective non-imaginal values of the legal system collide with the imaginal normalizing vision of the lead detective, Inspector Wexford.[62] Dogged and unimaginative, and so lacking Athena's divine potential for transformation, Wexford's deputy, Mike Burden, is fatally left in charge of a vulnerable suspect. He proves unable to see beyond the non-imaginal norms of the police.

Despite being warned by the suspect's Jungian analyst of his fragility, Burden continues to question the disturbed young man until the evidence no longer warrants it. Unfortunately, the suspect has become fixated with Burden on an imaginal level. In the analyst's terms, the suspect has made a transference. Now stalking the increasingly distraught Burden, the suspect takes all too literally the message that Burden is only interested in murderers. He kills to get Burden's attention.

All this occurs while Wexford is in hospital, having been wounded by a car bomb. Lying injured, he is visited with images that point him toward solving the case. Later he is able to counsel the horrified Burden over his inability to conceive of the imaginal way he was behaving to the disturbed suspect. What Wexford, Maisie, and Poirot prove is that Athena's normalizing vision *has to be able to encompass the imaginal* if the Furies are to find a place in the genre's container. Maisie "places" the Furies for her bereaved clients; Wexford teaches Burden they exist and must be treated with respect. And Poirot acknowledges their intolerable presence, leading to his encounter with Orestes haunting twelve persons who have killed.

It is time to assess Athena's normalizing gaze in terms of reason, persuasion, and necessity in a specific example of her strategic art. Can Athena in the mystery genre protect a city built by slaves?

Case History (20): *Dead Water* (2004) by Barbara Hambly[63]

> "To save you and Athene from the life I daily live… I will gladly play the despot…"
>
> "Then you don't remember slavery, clearly?"
>
> January said, "I remember it."
>
> —Barbara Hambly, *Dead Water*

If as Rinda West suggests, the rehabilitation of the Furies in *The Eumenides* provides a home for those who bear the shadow for the collective, for Athena's city, then the African-American population of Barbara Hambly's 1830s New Orleans set mysteries are still waiting for their place in a racist society. Not all African Americans of this period are slaves, since the Creole heritage of the city left a vulnerable community of free blacks. Musician, surgeon, and reluctant sleuth Benjamin January is one.

Married to Rose, who is trying to set up a school for girls to give them ambitions beyond becoming mistresses of rich white men, the Januarys' livelihood is threatened when the Bank of Louisiana is robbed. January, Rose, and their white friend, Hannibal Sefton, a consumptive opium addict, are commissioned to recover the gold. The mission means joining a steam ship travelling through country where even an African-American with freedom papers is liable to be kidnapped and sold.

Before they leave, however, January must confront the concentrated embodiment of the Furies in his own community, in the person of voodoo seer, Queen Régine, who may be poisoning one of his pupils. The meeting does not go well. Régine curses January, and later reappears just as the steamboat, Silver Moon, departs. January is convinced that she is on board, concealed in the darkness below.

> Could Queen Régine hear them, he wondered, down in the
> terrible dark of the hold? Was she able – he could not imagine
> how – to come out on deck, to move about silent in the night,
> seeking him like a vengeful ghost?[64]

The Furies, in their infliction of indiscriminate suffering as punishment without a sense of justice or restraint, are what the Black community is forced to endure in slavery, as Hambly's fine novels reveal. In the marginal and vulnerable position of January and his family, and the slaves of the surrounding cotton plantations, the Furies are present in the lack of reason, order, and justice in their lives.

The Furies are also what the whites fear in, and from, the slaves they daily oppress. Any hint of black armed resistance is punished with an "animal fury of terrors that the slave-holders would not even admit they felt."[65] Slavery incarnates the Furies; voodoo is one manifestation as a practice both outside and inside this cruel system. Drawn from West African inheritance, honed by desperation and pain, the Voodoo

Queens are an unpredictable and uncontrollable resource for the almost powerless. January doubts Regine's sanity. He finds his Catholicism small comfort against her curses.

Yet Queen Régine is not the only living vengeful ghost haunting the Silver Queen, with its cargo that includes two chained groups of slaves on their way to markets. The bank robber, Oliver Weems, is accompanied by Mrs. Fischer, a formidable woman who accuses Hannibal and January when Weems is killed. Another murder is then arranged when Hannibal is forced into a duel with pistols against Molloy, the drunken and vicious pilot. A raven croaks "like a vengeful ghost" as Molloy falls dead, apparently to the distraught Hannibal, but actually to a deadly sniper.[66] January begins to realize that the two groups of slaves are actually one group of humans for sale and another who are actually escaping in full sight. Disguised as slaves about to be sold, these men are seemingly the Furies' prey; in truth they are desperately collaborating with Athena in the garb of a white slave seller.

For Jubal Cain, the terrifying unleasher of Furies as slave dealer, is actually Judas Bredon, whose first name belies his principles and brave commitment to ending slavery. January eventually deduces that Bredon/Cain is actually working with Rodus, his chained "slave" who is brother to the much paler Thucydides, the ship's steward. Once Weems recognized the true identity of "Cain," his fate was sealed, as was Molloy's, whom Rodus shot from concealment instead of Hannibal. The conspirators of the Underground Railroad then poison Hannibal, so that Molloy's death will not hold up the Silver Moon, as any delay would imperil the escaping slaves. Thucydides assures January that they did not mean to kill the gentle Hannibal, who does narrowly recover. However, he admits that the Railroad is prepared to murder to secure their work.

In a world where the Furies are at large through the buying and selling of human cargo, trust is dangerous and necessary. Before January uncovers the real white Railroad leader, he meets Reverend Christmas, who pretends to rescue slaves while actually preying on runaways. Consumed by the tale of stolen gold, Christmas ambushes the Silver Moon, being prepared to kill and torture until he finds it. January tells him honestly that he now thinks the gold to be concealed is in New Orleans; Christmas does not believe him.

Set on fire, the steamboat sinks, but not before January returns to rescue Queen Régime in the dark hold. Instead of the old witch he finds a young black woman, Julie, a maid sold into harsher slavery during the voyage, and who had appeared to run away. Saving Julie at last brings a blessing to January and Rose, for Queen Régine comes to their house and removes the curse. Julie is Queen Régine's granddaughter and by virtue of her divine gift she has seen January rescue the girl. So here the Furies do indeed become "The Kindly Ones," as Queen Régine bestows a blessing, instead of a wasteland.

In the end January finds the stolen money, restoring the bank and his own, fragile security. However, he also discovers that Hannibal's name for his wife, Athene, indicates that his work to assuage the Furies is far from over. A knock at the door reveals that Judas Bredon did not die on the Silver Queen after all. He wants to recruit January and Rose for the Underground Railroad's work of rescuing slaves. Here January discovers his own conjunction of reason and necessity.

While there are Furies making a wasteland of psyche, spirit, and landscape through slavery, there must also be Athena seeking to convert them into the Kindly Ones. Slavery creates homelessness. Making people into slaves invokes the Furies, felt by the slaves every day, and feared by their oppressors. January and Rose will continue the work of Athena in seeking to instate justice and mercy in a land in which savagery is part of the social fabric. Slavery, Hambly shows, unleashes the Furies. There is no secure home, no safety for the city (of New Orleans or any other), while slavery exists.

Conclusion: Athena Stops the Furies

Detective fiction contains, holds, and transforms the Furies into the kindly ones. Without honoring these resident aliens no household can thrive. Mysteries accommodate the Furies by the sleuth standing in for their indiscriminate rage, just as Athena in *The Eumenides* insists on replacing their devastating fury with reason blended with necessity. The Furies are re-*placed*, and so are able to be *cultivated* in an ecological model of psyche.

So, in works ranging from Laurie R. King's *To Play the Fool* (1995)[67] to Carolyn Hart's *Mint Julep Murder* (1995),[68] unappeasable rage that could make a wasteland is instead re-woven into the containing powers of community and genre. In *To Play the Fool*, a destroyed life, that of pastor and Professor David Sawyer, is revealed in the enigmatic person of Brother Erasmus who is both homeless and minister to the homeless. Driven from his privileged life of status by feeling responsible for an unstable man killing five people, including Sawyer's son, Sawyer surrenders his will to wander among outcast.

Sleuth Kate Martinelli enlists scholarly and theological help to diagnose Erasmus as a holy Fool and uncover his trauma. By freeing Erasmus from accusations of murder, the Fool is freed to become the Kindly One among the indigent. Solving the new killing becomes a means of exposing and re-weaving past murders into the social fabric. *To Play the Fool* has a distinctively spiritual answer to the Furies; yet, in demonstrating that retreat into individual breakdown does not stop their persecution, just as Orestes discovered, it adheres to Athena's patronage of the mystery genre.

Here Erasmus/Orestes needs the Athenian work of legal and scholarly detectives, in Kate Martinelli for the police, and Professor Eve Whitlaw for theology. Together they uncover the criminal and traumatic past in order to find and make reason and necessity in Brother Erasmus's eventual return to the streets.

Far more cozy in tone, in *Mint Julep Murder*, blissfully wed Annie Darling has recklessly volunteered in a mystery writers conference by assisting five troubled winners of a coveted prize. When the manipulative publisher who will award the prize is poisoned, all five possess motives. Here the Furies are specters in the very real possibility of social breakdown with the group of suspects, who are all guilty of *something*, cooped up with desperate would-be authors and fans. The mystery conference stands for the possibility of indiscriminate and unceasing savagery that the community faces if these multiple sources of guilt and terror cannot be *solved* as in *dissolved*.

Fortunately, Annie proves a robust guardian of this particular afflicted city or conference. She manages to ally herself with the police so that she can perform the traditional Athenian ritual of rehearsing the crime in the uneasy presence of the suspects. Like Athena's hearing of Orestes and the Furies, Annie unearths two murderers by her strategically skillful handling of weaving narrating and questioning. By exposing the venal and the criminal among the suspects, the Furies have their place assured in the operation of justice. They can be honored by writers like Leah, no longer terrified that her husband will discover an affair, or Missy Sinclair, fearful of her unsavory past.

Athena is present in mysteries trickily. She is necessary to meet the Furies who rise up when murder may be left unsolved and unpunished. She normalizes and stabilizes by honoring the Father insofar as his power is mediated by the necessity of death. Athena is community made architectural in institutions that need to be rooted in necessity, reason, and persuasion.

Above all, Athena is in the detective re-*placing* the Furies. She does this in the ritual form of the genre itself. Reading mysteries is a ritual that invokes a number of god(dess)s, but especially Athena in affirming our social necessities. Mysteries offer Athena to the modern world, in containing and uniting through skillful weaving those psychic horrors that ferment murder, and those psychic rationalities that solve it in ways that make community sustainable and bearable.

For the next chapter, we consider a goddess more dedicated to joy, beauty, and Eros. Aphrodite is not to be ignored in the mysteries of living and in what may prematurely end our carnal existence.

CHAPTER SIX

THE MYSTERIES OF APHRODITE

Introduction

W here is dazzling Aphrodite in the pursuit of killers? Early chapters of this book explored the mythical kinship of detective fiction and the trickster.[1] Ubiquitous in world mythologies, the trickster hunts and is also hunted. He or she (for gender is as variable as a trickster's other qualities) stands for detective and villain, savior and devil. Indeed, the trickster is the mystery genre itself, tricking readers by inviting us to keep up with the sleuth. In doing so, the trickster genre arguably invokes an ancient structure of consciousness in both hunting and being hunted; the world of our hunter-gatherer remote ancestry. Such a possibility draws on C.G. Jung's notion of archetypes as inherited potentials for images and meaning.

How does Aphrodite, goddess of the beauty that inspires, in-spirits desire, enter this primal scene? Perhaps the often posited link between trickery and Eros offers a way in. For with the archetype of the trickster, the detective genre gains not only the theme of hunting, but also an inherited pattern of relations. The trickster is the drive to animate matter into signs, the consciousness of the Earth Goddess, and the chance of reviving her in modernity by connecting the making of meaning to the carnal body.

In my book, *Jung as a Writer* (2005), I suggested that in the mythical figure of the trickster Jung concealed the image of Earth Mother as a type of consciousness based on relating.[2] Relying on the pioneering work of Ann Baring and Jules Cashford, in *The Myth of the Goddess* (1991), where they suggest two types of creation myth standing for two founding structures of consciousness in Western modernity, I proposed that an attempt to re-align these mythic structures for the modern psyche is, in fact, the architecture of Jung's entire opus.[3] Earth Mother consciousness, explain Baring and Cashford, historically derives from pre-monotheistic animism referring to the Earth herself as sacred, as Mother to all beings. Prior to the division into two genders, Earth Mother is not dualist, not instating a fundamental cosmic structure of division or self versus Other. Consciousness *with her* is based on connection, relating, Eros.

Opposing Earth Mother consciousness is a monotheistic mythology installing a god who created this world as separate from himself. By that very structure, Sky Father installed dualism, two genders, spirit as separable from matter, and body as well as sexuality as other to the divine. Sky Father types of creation myth produce consciousness based on separation and division. Both types of consciousness, one structured on relating, the other on separation, are arguably necessary. Unfortunately Western modernity suffers from a long historical imbalance of its monotheism suppressing the connectivity and embodiment of the Goddess.

Here the Greek Pantheon of goddesses and gods can be regarded as a sophisticated and psychologically rich system of cultivating both Sky Father and Earth Mother consciousness as mutually life sustaining. After all, with Earth Mother being pre-gender division and Sky Father insisting upon duality, this stark opposition is ameliorated by the divine family of Zeus, Hera, and progeny with their habits of quarrelling, mating, and, yes, playing tricks.

Gender traits congealed by a society that overemphasizes Sky Father narratives were far more subtly displayed in these divinities. For example, Athena, as in the previous chapter, sides with the Father, to the extent of taking on many of Sky Father's qualities of rationality. In turn, ecstatic Dionysus is the god with greater allegiance to Earth Mother. For he embraces sexuality, body, as well as gender ambiguity in the sacred.

Aphrodite, as Ginette Paris astutely shows, parallels Dionysus in many aspects, yet her divine sexuality is a civilizing force rather than the brutality inherent in Dionysian rites.[4] Aphrodite is the Earth Mother refined into psychic complexity as feminine. Specifically, Aphrodite retains much of Earth Mother's resistance to dualism, in blessing same sex as well as heterosexual unions and, importantly, makes no distinction between nature and culture. This is vital in giving her *vitality*. She is the civilizing beauty of gardens as well as the desire evoked by humans naked and adorned. As Christine Downing notes, Aphrodite is the sexuality of wolves as well as humans.[5]

As goddess, Aphrodite's divinity is neither safe nor comfortable. If she initiates into mysteries of the body in carnal connection, she also sends us to the underworld, through loss or the pain of intimacy betrayed or broken.[6]

Above all, Aphrodite is necessary for creation through procreation. She bestows the beauty that inspires erotic desire.[7] She is connection through sexuality as initiation into the sacred. In this she embodies and enacts an Earth Mother who is matter herself, the mater who is matter as sacred, not other to the disembodied divine. Therefore, Aphrodite is descendant of Earth Mother in the carnal body and in sexual union as sacred and therefore meaning making. She too animates the matter of desire into meaning, into signs. If she is implicated in the body that knows (erotically), or the body that must be cultivated into clues, then Aphrodite is indeed necessary to the sleuth invoked in women's mysteries.

Aphrodite's Body and the Sleuth

> "That one, a homeless man was fishing, hooked up and pulled an arm out of the water…"
>
> Great restraint, Mikey. He had resisted the temptation to tell Lunetta that he has christened that victim, "Venus." A one-armed Italian woman in a cement overcoat didn't lend herself to any appellation except Venus de Milo.[8]

While the erotic body can be an organ of knowing for the detective, as we will see, the body is also the primal site of the mystery. In fact, a corpse might be doubly crafted for Aphrodite. It is both corporeal matter requiring animation by the sleuth's connecting to its injuries as meaningful, and, additionally, may have fallen foul of Aphrodite's powers more directly. Its death may represent the darkness of this frequently dazzling goddess.

With lust and lucre as frequent motives for murder, a corpse is likely to be scrutinized for sexual clues as to its demise. Aphrodite as inspiration for sexual desire can also be goddess of death, as her affair with Ares, god of war, indicates. Sex can be deadly. Aphrodite's bright divinity has a shadow aspect of the demonic in such underworlds as sex addiction, pornography, domestic violence, rape, and the desperation engendered in those abandoned.

Mythically, the fate of those who reject Aphrodite is a warning of sexuality's potency in human agency. Hippolytus, moon and Artemis worshipper, spurns Aphrodite by turning away his stepmother Phaedra's advances. Falsely accused by her, Hippolytus dies at the hand of his father, Theseus, a murder pre-figuring many domestic sacrifices to Aphrodite. Those who refuse sexual desire refuse Aphrodite. They do not fare well because they are ignoring an aspect of the psyche that is essential to life. On the other hand, Aphrodite alone is no haven. In choosing Aphrodite as the fairest who will receive the golden apple, Paris accepts her offer of the most beautiful woman in the world, Helen. The consequence is ten years of war and the destruction of his city of Troy.

Aphrodite rejected or despised wreaks vengeance in sexually driving us into the dark. Alone, Aphrodite offers neither peace nor permanence.

So other goddesses and gods are essential to mediate and weave Aphrodite's radiance into something more nurturing for the psyche and the community to survive. Aphrodite's beauty requires Athena's reason and necessity, and the wild ecology of Artemis, as well as Hestian centering, in order for a relationship to endure beyond this goddess's native spontaneity and irresistibility.

And yet Aphrodite inhabits mysteries because her dazzling erotic energy is fundamental to desires that both destroy and restore. The goddess who kills for, or because of, sexual love is also the drive to know the beloved in healing, restorative Eros. Since solving murder is often motivated by the desire to redeem individuals or a community, Aphrodite inspires the sleuth as much as she may tragically or demonically inhabit the sexuality that kills. Here Aphrodite is the drive to animate the dead matter of the corpse into meaning, clues, the connection of completing an-other's story that the sleuth embodies. Aphrodite is the Eros of knowing that drives detecting.

Therefore, above all, Aphrodite gives us embodiment as erotic knowing and sexual connecting. She is the orgasmic energy of life that intimately knows extinction. For Aphrodite, orgasm is sacred, and if not respected, it can be an initiation into hell.

So yet another precise aspect of this goddess in mysteries is their pivotal axis around murder itself. Aphrodite in detective fiction reminds us that a murdered person is a violated body. Whether or not sex was part of the murder, the corpse has suffered death inflicted by another, rather than life ending according to the necessity of nature. Corpses are not beautiful, or at least are destined to lose their appeal in decay. Although capable of exciting desire in some who have a torturous connection to this goddess, the dead body is of itself an offence to Aphrodite in its premature loss of carnal pleasure. Hence the re-reading of the body marked by unnatural death is to take violated matter, such as blood on a floor, and animate it into meaning as a service to this goddess.

These signs, the blood as image of biological knowing, become more imbued with significance as the story progresses; for example, telling also of the manner of death. The blood as sign becomes a Jungian symbol in offering what is known, what is imagined, and what is not yet known or knowable about the murdered person. The signs of the body are here resurrected in the imaginations of sleuth and reader.

To the extent that we mobilize the body as signs into a corporeal, desiring, erotic being, we invoke Aphrodite to know, to revivify, or revive by our embodied psyche. In this sense, the goddess, offended by brutal ending of sexuality inspired by beauty, is appeased by mysteries. These fictions are themselves a work of Aphrodite countering a brutality of the body more liable to belong to the violent ecstasies of Dionysus. Just as there is Athena making a place for the Furies in mysteries, so there is also Aphrodite Eros countering and limiting Dionysian excess.

Mysteries, with their roots in the trickster as figuration of Earth Mother consciousness, are indeed Athenian in weaving and containing communal trauma. They are also modes of Artemisian hunting and Hestian homemaking and defending. Yet mysteries are also Aphrodite's redemption in a world where she has been neglected. Perceptible in that which drives killers is Aphrodite scorned, ignored, despised, or repressed. As the energy behind resurrecting the carnal body of the deceased as a breathing, beauty-inspired erotic being, Aphrodite's animation of matter into clues is the mystery genre come to life in the imagination of the reader.

It is time to look at a detective who is forced to explore Aphrodite in herself, in the crime, and as impersonated in a world sadly lacking her ability to give joy.

Case History (21): *Murder by Mocha* (2011) by Cleo Coyle[9]

> Alicia's boss was an enigmatic businesswoman known
> only by the name Aphrodite. Just a few years ago she'd
> started a Web site called Aphrodite's Village Online.
> —Cleo Coyle, *Murder by Mocha*

Manager of an historic New York coffee house, Clare Cosi is unusually accustomed to dead bodies. Yet she fails to notice that the particular bloody corpse in bed next to client, Alicia, is, in fact, still alive. Alicia works for Aphrodite, a ruthless businesswoman who encourages competition so cutthroat amongst her employees that at the launch of Mocha Magic, a so-called aphrodisiac-enabled instant coffee with chocolate, one unlucky worker is actually stabbed. Clare's sleuthing instincts are aroused, along with her desire for lover/cop Mike Quinn. It seems that Mocha Magic may be more potent than advertised! Has Clare been conned into a business venture that blends her coffee with illegal drugs?

Murder by Mocha offers in the recluse Aphrodite the commercial exploitation of myth that proves to be an invocation of divine and demonic energies. Apparently a successful marketing ploy, Aphrodite's "Village Online," with its various "Temples" or divisions, is intensifying what James Hillman calls a modern incarnation of the goddess in the seduction of money in hyper-capitalism.[10] To emphasize the point, Mocha Magic's supposedly fake addictive properties turn out to be real. The laced chocolate and coffee inspires love-making at the launch party that serves to distract attention from murder. Addiction to "love" kills. Mocha Magic is a metonym for the drug-like quality of shopping for sexual satisfaction.

So, if simulated Aphrodite has real aspects of the goddess in stimulating desire through exploiting myth in marketing, Clare herself, as a detective, also encounters the goddess. Unable to make much progress with Alicia, Clare cannot understand why Madame, her ex mother-in-law and employer, insists on offering unconditional support to a murder suspect even if she is a friend. Then a dangerous looking older man turns up at the Village Blend coffeehouse looking for Madame. Quinn reveals that this figure is Cormac, a corrupt ex-cop and killer, for whom the apparently law-abiding Madame went to jail to protect.

Clare's love for Madame ensures she listens to her side of the story, which proves to be one of Aphrodite's most poignant encounters. Widowed Madame met Cormac the cop and they became lovers, planning to marry. Unfortunately Cormac discovered corruption and murder within his police squad and was himself framed before he could bring the truth to light. Fleeing an endemically unjust legal system, Cormac left Madame, pregnant with a daughter. She went to jail for refusing to reveal his whereabouts. Suffering Aphrodite's abandonment, Madame also lost the baby, whom she had planned to call Clare.

Not long afterwards, Madame's wayward son Matteo returned from Europe with a pregnant bride, Clare herself. A daughter is restored to Madame, one who becomes the sleuthing ex-daughter-in-law of these coffeehouse mysteries. This sleuth, through love, believes Madame and trusts her Aphrodite enlightened knowledge of Cormac, even though there is no material proof to support the word of his lover. Will Mike

Quinn, tasked with arresting Cormac, believe *his* lover, Clare, or the "towering command structure he'd trusted for his entire career"?[11] Such a phallic description of the police also suggests a saturnine rigidity and hierarchy inimical to Aphrodite.

Fortunately, Mike has the trickster qualities of a "wily Odysseus," suggesting Aphroditic (trickster-goddess) qualities flourishing through his union with Clare.[12] While he, Odysseus-like, quests for evidence to bring Cormac "home," Clare and Madame discover entrepreneur Aphrodite's origins are one Thelma Vale Pixley, the author of a notorious and salacious myth-soaked dissertation at college. Striving to act out pagan values, Thelma seduced a married professor whose wife promptly shot him dead in front of witnesses. Given that the witnesses, Alicia and Sherri, testified against the wife, who died in prison, Clare wonders if surviving child, Olympia, is out for revenge.

After Sherri is killed and attempts on Alicia's life multiply, Clare is distracted by another Aphrodite-inspired crime. Her new barista, Nancy, has drugged fellow barista, Dante, in a forlorn effort to seduce him. Dante's stay in the narcotic-induced underworld is happily brief; Clare follows Nancy to the chocolate factory where Mocha Magic is made. There she discovers a demonic version of Aphrodite rising from the sea. Olympia, who had been an employee of Aphrodite under the name of Daphne, had previously faked her own death for the second time. She threw a clothed ice sculpture of Venus de Milo into the Hudson river.

Drowning an ice figure in water might indicate the ephemeral quality of Aphrodite's attention, but it also ensures that bystanders can testify to seeing a clothed figure visibly sink. Now risen from the waves, Olympia is determined to kill those she holds responsible for the death of her mother who, in turn, had murdered a faithless husband. Thelma/Aphrodite's attempt to live out the goddess as desire without being limited by marriage (just as the goddess was not restricted by *her* marriage) finally proves self-defeating. After all, exclusive attention to one goddess is liable not to end well. Olympia shoots modern Aphrodite dead, while Clare and Nancy, with marked Hestian qualities between them, survive to love again.

In *Murder by Mocha* Aphrodite manifests darkly as well as in the joyful surrender of Clare, Mike, and formerly Madame to sexual passion. As a lover, Clare uses Aphrodite's knowing to support Madame and influence Mike, the cop. Carnal bodily knowledge is part of their quest for truth. Yet a world without reverence for Aphrodite as the sacred in sexuality is also subject to her revenge, in addiction to drugs as addictive as shopping, as chemical highs, and to money as a substitute for love. It is time to look further at those mythical patterns that incarnate such a fascinating and destabilizing goddess.

Inherited Pattern of Relating (1): Birth from the Sea

> Like the goddess, we have gone far from the original
> ocean; and, like her, we must renew our spirit in the
> sacred bath, by fusion with sexual wetness.
> —Ginette Paris, *Pagan Meditations*

As Ginette Paris demonstrates in *Pagan Meditations* (1986), some legends of the birth of Aphrodite pre-date patriarchal inscription. The goddess emerges from the sea and returns periodically to renew her sacred virginity. Here "virgin" signals her autonomy, her freedom from restrictive ties to the masculine. She is whole, alone, and her integrity is not compromised by her ecstatic sexuality. So Aphrodite offers a corporeal feminine arising from the sea of unconsciousness and one capable of independently renewing herself by re-immersion.

In this sense, the sleuth invoking Aphrodite is embodying knowledge via descent into the unconscious and return. Something primal and innate to mysteries is imaged in this process of complete immersion of body and psyche, in order for "knowing" to emerge. These detectives do indeed commit their bodies to the waters of unknowing, plunging into a shadowy realm wet with blood, tears, and the excretions of a typical murder scene. In her quest for truth, the detective risks body and mind, often to the dissolution of death, before she can both be Aphrodite rising from the waves (of unconsciousness or unknowing), and simultaneously summon her.

The culprit is invoked by the detective, not merely identified by her. For it takes the bodily and psychic descent of the sleuth into the dark waters for one of its creatures, a shadowy suspect, to be incarnated through the animation of matter into clues, as the perpetrator. In this sense, the murderer is indicted as dark Aphrodite, a goddess whose addictive qualities of sexual lust, or lust for lucre, are acted out in the killer.

For example, in *Toxic Shock* (1988), V. I. Warshawski conducts exhaustive rounds of investigation, initially yielding little information about the poisoning of workers in a Chicago chemical plant.[13] Finally the truth arises from a sea of evasions, that the factory's doctor was aware of its lethal effects. Together with the doctor's sister, V. I. forces him to recognize that his dependence upon money and power seduced him into collaborating with corporate murder.

Paretsky's indomitable sleuth, in one way so psychically of fierce Artemis, is in fact here deeply Aphrodite in her methods. For V. I, like Kinsey Milhone, *immerses* herself to the point of risking corporeal dissolution in a complex underworld. For V.I. Hades is the poisonous waters of capitalism and its urban shadows, and for Kinsey, the equally perilous currents of her family. These sleuths are Aphrodite, in their ability to sink into a mass of confusing signs around a crime and animate them through body and psyche. What is formerly a cruel sea of not knowing begins, by the animation of the goddess, to condense matter into clues until a solution emerges.

Yet Aphrodite is also renowned for a more specific birth story with strategic patriarchal interventions. Ouranos, the Sky, oppresses Gaia, the Earth, who encourages her children to rebel. After Cronos castrates his father, Ouranos' semen is scattered on the sea as foam.[14]

And from this white foam was born the beautiful Aphrodite.[15]

Born in the sea from the father's genitals, perhaps this version of Aphrodite is visible in the sleuth's often erotic relationship with the cops. Kinsey Milhone actually was a cop before emerging from that patriarchal institution as a Private Investigator. Arguably, she re-immerses herself in this origin myth by her on-off, sexual relationship with a married cop in Sue Grafton's early alphabet books.

V. I. Warshawski, Sharon McCone, Stephanie Plum, Hannah Swensen, Goldy Schultz, Clare Cosi, and others all have romantic relationships with cops that facilitate their vocations as detectives. Alexandra Cooper, District Attorney in charge of sex crimes, has a particularly intimate, so far non-consummated, erotic relationship with homicide detective Mike Chapman. Their friendship is indivisible from their detecting partnership, perhaps made possible by the third of this quest-hero trio, the fatherly cop, Mercer.

For these fictional detectives, who span any perceived "hardboiled" and "cozy" divide, the solution *emerges* from the sleuth's immersion in the sea of unconscious unknowing, precipitated by the messy, slippery, and liquefying stuff of the corpse. Yet Stephanie, Kinsey, Hannah, V. I., etc., are all *inseminated* with vital ingredients for truth by their erotic links to the police. These sleuths are Aphrodite, a generative divine feminine figure, in that they contribute creativity to the seminal fluid of police procedures. Without these unofficial questing characters, dark Aphrodite would not arise from this sea.

It is time to take a closer look at Aphrodite's disastrous potentials for human desire.

Dark Aphrodite's Lusts: consumerism, celebrity, addiction to money and sex

> At first, it has been for the attention, and then to get
> the job on the cartoon show. But the real irony was
> that he'd killed a man, and had been willing to kill
> more, just so people would love him.
> —Toni L. P. Kelner, *Blast from the Past*

In the new series from Toni L. P. Kelner,[16] the lure of celebrity, the seduction of fame, fans, and its presumed riches, is a stimulus to crime. We follow Tilda Harper, struggling media journalist and writer of "Where Are They Now," articles on stars of TV series past. In a world where Aphrodite does not get adequate vehicles of expression for her sacred sexuality, she may appear in addictive potency as pornography, celebrity, and consumerism. Just as, on the one hand, Aphrodite's beauty inspiring desire may dazzle in the person of a great film star, on the other, in a degraded era of sex divorced from the sacred, the goddess may drive a desperation for fame that is prepared to kill.

In *Blast from the Past*, a former stuntman disguises himself as an on-screen character, a monster, in a bid to become Eros to the viewer's enraptured Psyche. In the myth, Psyche fears she has married a monster, but discovers in Eros his mother's (Aphrodite's) desire-inspiring beauty. Here the monster is more than skin deep. As the killer tries to impersonate Aphrodite's beauty, he succeeds only in invoking her annihilating anger. (See section on "Eros and Psyche" below).

In consumerism as the dark side of Aphrodite's penchant for adornment, Hillman and Paris emphasize different aspects. Hillman insists upon repressed Aphrodite's "pink madness" of pornography and seduction in consumerism,[17] while Paris explores the positive generative aspect of Aphrodite's exquisite clothes, her beauty enhanced by the very gold that is also used to buy love.[18]

In mysteries, the inseparability of money and sex, or love, motivates many a murderer. These range from the sex crimes in monied settings, as in Alex Cooper's New York, to Clare Cosi's encounters in the same city with those driven to revenge over the cruel severing of intimacy. Yet Clare also knows how Aphrodite, in her beauty enhanced by fine dresses, can be indispensable to the (divine) pursuit of knowing though the carnal body.

> If I ever write a manual on how to be an amateur detective, I will add a chapter on one of the most important assets any investigator can have – an impeccably dressed elderly women who[se] ... presence is so imperious, so gracious, almost no one will question her motives or rudely ask about her business.[19]

Madame, the ex-mother-in-law of sleuth, Clare Cosi, performs many Aphroditean functions, not the least of which is her commanding personality made more enchanting by clothes that combine taste and expense in service to the goddess. Less cultivated than Madame, Clare herself uses adornment to intensify her erotic attractiveness when sleuthing. In less decorous circumstances, Stephanie Plum is also prepared to dress up her sexuality, when Ranger, her enigmatic and dangerous suitor, requires a quarry to be enticed into his custody.

While sexuality can easily be deployed in the service of entrapment, on behalf of detectives or criminals, Stephanie Plum does have a distinct relationship with Aphrodite in terms of consumer desire. Often accompanied by black "ex-ho'," the fabulous Lula, both women are easily seduced by cake and fashion. In *Four to Score* (1999), with Lula and transvestite, Sally, when engaged in bounty hunting, shopping is indispensable to success.[20]

> It was a little after one when we got to the park.
>
> "Those shoes make all the difference," Lula said, staring down at my new shoes. "Didn't I tell you those shoes were the shit?"

"Slut shoes," Sally said. "Retro fucking slut."

Great. Just what I needed, another pair of retro slut shoes – and an extra $74 on my Macy's charge card.[21]

Like so many detectives in woman-authored mysteries, Stephanie Plum narrates her own adventures. Here the first person perspective introduces Aphrodite's erotic consumerism, and even makes it more attractive through humor. Aphrodite is seductive because she delivers on a good time. Through Stephanie's erotic fascination with clothes and sweet food, as well as fantastic sex with cop, Joe Morelli (and even, occasionally enjoying the mythical prowess of Ranger), we, as seduced readers, are imbued with her divine charm.

Now let's consider a detective who fears that pregnancy has severed her relationship with sex as sacred mystery.

Case History (22): *The Girl in the Green Raincoat* (2008), by Laura Lippman[22]

Tess Monaghan, heavily pregnant and suffering from high blood pressure, is confined to a chair by a window for the sake of her infant. Previously determined not to be a victim of love, she had kept boyfriend, Crow, at a distance, even initiating several break ups. Now she no longer feels in control because she isn't. Huge and dependent upon others for meals, news, and any connection to her life as a Private Investigator, she has lost all sense of Aphrodite's charms. Nor does she consider herself maternal and is charged with anxiety about becoming a mother. She refuses to accept gifts for the baby or to agree to preparations in their home before the child is born.

So Tess is stuck. Physically confined, she is also scared of what she cannot control—her pregnancy and changing relations to Crow. Aphrodite, patroness of her previous "virginal" autonomy and erotic sense of adventure, has fulfilled her fidelity to transient relationships by vacating Tess's relations with Crow. She now longer feels that "spark" between them. Desperate to find distraction, Tess, gazing from her window, notices that a woman in a green raincoat fails to appear for her routine walk with a dog.

Enlisting Crow and her best friend, the aristocratic Whitney, to make inquiries, Tess discovers that the woman, Carole, has disappeared.

She is married to Epstein, a man with an unfortunate history of two previous dead wives. When Tess finds out that Epstein dated Carole's sister, who also died, she becomes convinced that here is a likely case of Aphrodite's charms turned deadly. Epstein is surely a serial seducer and killer. The trapped P. I. convinces Crow and Whitney to lure key witnesses to visit her, in a novel punctuated with conversations about love and marriage.

Deserted by Aphrodite herself, Tess now dreads Crow leaving her and notes a conspicuous absence of conversations about marriage. She asks the homicide cop from Epstein's first wife's shooting about love and commitment.

> "Did you know your wife was the one, the moment you met her?" she asked. "Or did it creep up on you?" Her relationship with Crow fell in the latter camp, and she couldn't help thinking there was something special about the thunderbolt school of love.[23]

Tess's hormonal anxieties perhaps confuse Aphrodite and Zeus here. The divine patriarch with his characteristic lightning bolts was in the habit of assaulting women with an unpredictability that was sometimes fatal. By contrast, Aphrodite's seductiveness is more subtle, and can manifest over time as well as be instantaneous. On the other hand, Tess realizes here a connection between detecting and Aphrodite as a complexity of knowing in the psyche. The cop was "a good murder police" in getting people to confide in him.[24]

Love is a universal invitation to the sacred in touching deep being. While it is a likely ingredient in crime, its very erotic connectivity is a channel between people for sharing, including the sharing that enables truth to emerge from the sea of unconscious unknowing. The birth of Aphrodite is the birth of knowing where her charm seduces details into animation, pattern, and the realization of concealed motives.

Aphrodite's presence in same sex love comes to Tess's confinement in the persons of Liz and Beth, lesbian adoptive mothers of Chinese born May. Crow, while *not* discussing marriage with Tess, has given a family heirloom ring to Lloyd, a black teenager he has been helping away from an impoverished and dangerous childhood. Lloyd is told to give it to the woman he is to marry, and so promptly becomes engaged to May while both are still teenagers. May's white mothers are sanguine about a black boyfriend, yet are horrified by the prospect of a career blighting marriage.

A lunch takes place in Tess's room, in which Aphrodite is celebrated while marriage is at first considered by most of those present as not hospitable to her charms. The exception is Liz, who wants to get married; Beth does not. May is distressed by their fighting. Neither of her mothers wants May to commit so young. Yet they are a happy couple who lost ten years together because of social attitudes to their sexuality. When asked if they regret not being together from about the age of their daughter, both reply, "I do."[25] It is Lloyd who points out that they have uttered the key words in the marriage ceremony.

Here Aphrodite and marriage are momentarily united in the love of Liz and Beth as they express an Aphrodite sorrow at parting due to cruel external pressures. They find a compromise on a marriage ceremony, restoring Aphrodite's pleasures in that union, while Lloyd and May agree to wait.

Meanwhile, Tess and Whitney decide to try Aphrodite's seductive wiles as a detecting tool in pursuing the suspect, Epstein. Whitney becomes his girlfriend, then finds herself unable to tell Tess that she actually enjoys the dates. However, when Epstein takes her to a "sacred place," and it proves to be the graves of his first two wives, Whitney discovers that his confession is not what she and Tess expected.

For at the very same moment, Carole, apparently not dead, visits Tess with a Taser, a non-lethal stun gun that enables her to paralyze her victims before disposing of them. She is killer not victim. Tess's only hope is her regular delivery of meals, yet supper is late! Finally, it is Carole's abandoned dog, Dempsey, taken in by Tess who has two other hounds, who comes to the rescue. Carole is dark Aphrodite who kills for money, jewels, and fine clothes. Fortunately for Tess, in bonding to neurotic Dempsey she has also filled her Artemis potential. A combination of dogs and the belated arrival of Lloyd save Tess's life. Unfortunately the brush with death provokes premature labor. Born early, Carla Scout goes into neonatal intensive care and Tess finds she does do instantaneous love after all, at least for her baby.

Consumed with guilt, Tess keeps asking if she is responsible for the baby's problems, a question repeatedly answered in the negative. Tess did nothing to cause the placenta to fail, which would have required the baby to be delivered early anyway. What is more terrifying for both parents is that they cannot wholly protect their child. In the hospital, the nurses are the "goddesses," but even they cannot defeat necessity of death in the cosmos, as we saw in the previous chapter on Athena.

Baby Carla gets stronger and at last can go home. It is Tess and Crow who have not recovered from her birth. They have to get over it, says the doctor: "[w]elcome to parenthood."[26] What Tess has already got over is doubt of Crow's love. "I guess we can get married now," he says, revealing that he saved the best family ring for Tess.[27] When facing death from Carole's Taser, Tess had spoken to Crow on the phone of her love for him. In the story of Tess and Crow, *The Girl in the Green Raincoat* invokes Aphrodite in the unromantic dimensions of erotic love.

At first Tess risks Aphrodite's relative lack of maternal qualities, a shady aspect of the goddess compounded by Carole's seduction of a rich man to feed her obsessions with clothes and money. Yet by investigating whether Aphrodite can inhabit long term relationships, as well as enduring an Aphroditean underworld experience of fear of abandonment, Tess discovers her deep bond with Crow, especially when faced with her death of their daughter. She finally learns that Aphrodite can be a gift *from* the beloved.

Her quest for Aphrodite when pregnant leads this procreating sleuth into solving several murders. Tess learns that erotic love is a form of knowing she can trust, even though this goddess remains tricky and possessed of darkly seductive powers.

Inherited Pattern of Relating (2): The Golden Apple

When Eris, goddess of discord, is the only Olympian not invited to the marriage of Pelops and Themis, she sends instead a beautiful golden apple "for the fairest." Zeus insists that mortal Paris choose between Hera, Athena, and Aphrodite. Winning the apple, Aphrodite promises Paris the most beautiful woman in the world, who happens to be Helen, the wife of Menelaus of Sparta. So begins the Trojan War, ending in the destruction of that legendary city.[28]

Aphrodite is not only associated with war through her lover, Ares. Yet the fateful story of the golden apple is more than the psychic link between eroticism and death, or sexuality and violence in its darkest form, as rape, or sex driven killing. Rather, the story of the golden apple is a myth about the disaster of ignoring less attractive parts of the psyche, such as Eris represents. Omitting Eris in the hope of a quieter wedding leads to a choice between goddesses, one that elevates Aphrodite to unusual prominence in human affairs. Running counter to the protean properties of the polytheistic psyche, Paris's preference for one goddess over others engenders war.

The story of the golden apple suggests that all goddesses and gods have their place and must be honored. Ignoring, repressing, or inflating one aspect of the psyche over others can lead to disaster. In particular, we see the mistake of gods out of balance in mysteries. These novels expose the deep psychological fissures in detectives, victims, and in those whose promotion or repression of a goddess such as Aphrodite drives them to kill. Murder is here a metonym for war. For choosing one goddess over the restraining presence of others serves to let loose Eris.

Hence, in *The Ides of April* (2013), informer Albia, daughter of Lindsey Davis's long term sleuth, Marcus Didius Falco, misidentifies her killer for too long because she is erotically attracted to him.[29] Ironically, she chooses Aphrodite because she is also *under*valuing the goddess as a widow still mourning lost love. For her own depths to be stirred enough to see *her* grief, she has to re-experience Aphrodite's underworld in witnessing that of a victim's widow. In fact, the end of the novel suggests that Albia has been evading Aphrodite as well as allowing herself to be beguiled by her. She has been refusing to acknowledge her erotic attraction to a fellow investigator, until she is forced to re-examine her sexuality as a potentially powerful form of knowledge.

In Sara Paretsky's *Bitter Medicine* (1987), a victim is tragically caught up in her family's determination to apprentice her to Athena, rather than Aphrodite.[30] Consuelo, a talented daughter of a poor family, dies in childbirth at age sixteen. As V. I. Warshawski told her, she allowed herself to get pregnant to escape the weight of family expectations. V. I. will also discover that Consuelo died due to hospital malpractice, under the all too frequent preference of greed over other, more humane, considerations.

A very different murder victim is Ronni Ward, in Joanne Fluke's Lake Eden mystery, *Cream Puff Murder* (2009).[31] An accomplished acolyte of Aphrodite, Ronni's generous sexual favors, including a fling with one of Hannah Swensen's suitors, Mike, the cop, offers a range of possibilities for the detecting cookie shop owner. Another victim enjoying an over-exclusive connection to Aphrodite is Denise Caxton, in Linda Fairstein's *Cold Hit* (1999).[32] Trying to extricate herself from a controlling marriage to a rich art collector, Denise has several lovers, one of whom persuades her to dabble in the underworld of art theft. Alexandra Cooper is drawn into the case when Denise's body is recovered from the river with evidence that she was raped before death.

Neglecting one or more fundamental aspects of the psyche is deadly to well-being and fatal to health. Aphrodite's role in the Trojan War demonstrates that this goddess is not a guardian of human community, like Athena; nor is she a maker of familial safety, like matriarchal Hera, or tends hearth fires as does Hestia. If worshipped singly, Aphrodite becomes dark with lust, dividing families and provoking murder. Indeed, so dangerous is neglect or over-emphasis of this goddess that lack of respect for her fuels the problems of Psyche herself, in the myth of Eros and Psyche.

Inherited Patterns of Relations (3): Eros and Psyche— Marrying Monsters

Narrated in the second century CE Latin novel, *The Golden Ass* by Apuleius, the story of Eros and Psyche is also found in Greek art from six centuries earlier.[33] The story begins with Psyche as the fairest and youngest of three sisters. Her overwhelming beauty causes her admirers to forget to pay proper respect to Aphrodite. Angered, the goddess gets Apollo to tell Psyche's father that her husband will not be human. Fearing a monster, Psyche is taken to a rock in funeral attire.

However, instead of being killed she finds herself in a beautiful house with a husband, Eros, son of Aphrodite. He only visits her in darkness and forbids her to look at him. In *Amor and Psyche* (1956), Erich Neumann famously interpreted the tale of Eros and Psyche as the individuation of the feminine.[34] Psyche begins her relationship with Eros in darkness, the unconscious. She is forbidden to *know*, to become conscious.

For the female-authored sleuth, Psyche's marriage to an unknown being, reputed to be monstrous, can stand for the beginning of a case of unknown dimensions and horrors. Some offense to Aphrodite, Psyche's suitors' lack of respect, the premature ending of carnal life, has activated the goddess's rage. In *A Drink of Deadly Wine* (1991) by Kate Charles, David Middleton Brown is summoned by a lover who long ago abandoned him.[35] Gabriel, now a married Archdeacon in the Church of England, is receiving threatening letters about the mysterious death of another ex-lover of his. David is the only person Gabriel can trust, despite having blighted David's life by his desertion.

Fortunately, Emily, Gabriel's understanding wife, has a best friend, Lucy, who immediately forms a bond with David. *A Drink of Deadly Wine* circles around the price and necessity of becoming conscious of love and sexuality. David has to begin to pay homage to Aphrodite, whom he has neglected since being rejected by Gabriel. Acknowledging a growing attraction to Lucy enables a closeness to her that helps them see that the malicious letters derive from an unconscious source. They were sent by a woman whose terminal illness gives her periods of darkness in which her rage and grief are as unbounded as that of an angry goddess.

Of course, Psyche in the original myth is persuaded by her jealous sisters to look upon her sleeping husband with the aid of a lamp. Waking, he immediately flies away, leaving her bereft, but not before she sees his immortal beauty. Echoing Psyche's fears, it is not uncommon in mysteries for one to discover oneself married to a monster, in that a trusted intimate is found to be deceitful. In Sara Paretsky's *Burn Marks* (1990),[36] V.I. Warshawski discovers she is dating a corrupt cop; while in Linda Fairstein's *Night Watch* (2012), Alex Cooper endures a thorough experience of Psyche distrust of Eros, when uncovering *circumstantial* evidence that her lover may be involved in fraud and violent death.[37]

Perhaps more innate is the mythical underpinning of marrying monsters in the detecting of Cleo Coyle's Clare Cosi and Diane Mott Davidson's Goldy Schultz. For these first husbands *are* monsters. Clare marries attractive playboy Matteo, son of renowned New York coffeehouse owner, "Madame." Struggling to ignore Matteo's monstrous infidelity and coke habit, Clare partly reverses the myth. She finally acknowledges Matteo's inability to change his endemic monster aspects, and divorces him. Unlike Psyche, Clare manages to become conscious that Matteo really is a monster, despite his considerable "erotic" charms. Now Clare, gingerly, approaches Eros again and begins a relationship with homicide cop, Mike.

Madame appears also to be reversing the myth, because unlike Aphrodite with a daughter-in-law, she has always welcomed Clare with love. And yet she starts to make life tricky for Clare in her new relationship, attempting to manipulate her son and Clare into remarriage. On the other hand, Matteo and Clare have already completed the story of Eros and Psyche by giving birth to a daughter named Joy, exactly as in the myth. Perhaps this is why Madame, who is enthusiastic to join Clare in tracking (or becoming conscious of), killers in their midst, gradually starts to let go.

In a very different way from Clare, Goldy also marries a monster: serial violent abuser, John Richard Korman, known as the Jerk to both his ex-wives. Several mysteries pivot on the struggle of Goldy to leave the Jerk and his subsequent infliction of ordeals upon her, which brings us to the latter part of Psyche's story. Yet before looking at the arduous tasks imposed by Aphrodite on abandoned Psyche, it is worth looking at a particular example of a sleuth tangling with this capricious goddess, Maisie Dobbs, authored by Jacqueline Winspear.

Case History (23): *Elegy for Eddie* (2012) by Jacqueline Winspear[38]

"I think… we've probably helped each other realize
that we still know how to love."

"I want to learn how to just let things happen sometimes."
—Jacqueline Winspear, *Elegy for Eddie*

Maisie Dobbs, private inquiry agent in 1930s London, realizes at the conclusion of *Elegy for Eddie* that she lacks in her life some of Aphrodite's spontaneity and life-affirming passion. What the novel also shows is that British society in that period was maimed by the devastating effects of World War One and stunned into unfeeling by the looming prospect of another conflagration. It is not only Maisie who suffers from Ares insufficiently allied to Aphrodite.

Eddie was a poor man of limited intellect but with an amazing gift for working with horses. Beloved by his community, Maisie is approached when he dies mysteriously in a paper factory owned by newspaper magnate, Lord Otterburn. Born in a stable to a single mother, Eddie's mythical origins mark him as "special," just as his unusual gentleness and innocence also set him apart. Maisie discovers that his gifts of Artemisian oneness with horses do indeed mark him out to people of power.[39] What the powerful fail to notice is the difference in Eddie, who is said to have his thoughts in his heart and not in his head.

While Eddie is enlisted by Otterburn to tend to his horses, an ambitious journalist suspicious of Otterburn's immense political influence sees Eddie as a means to obtain secrets. When another man, long jealous of Eddie, is employed to warn him off, death ensues for Eddie, the manipulating journalist, and finally, the bully. With the latter two deaths passed off as suicide, Maisie refuses to accept such convenient labels. "Suicide" does not tell her what she needs to know, she tells her lover, James Compton, who seems to be drawn half-willingly into Otterburn's far reaching plans.[40]

Here Maisie is like Psyche in needing to know the truth about Eros, about knowing in the body and in love, even though circumstances and powerful people warn her to stop such a search. At this point, as in other places, Maisie asserts the value of imagination in service of the truth, as opposed to the convenient fictions surrounding Eddie's demise.[41] It was Maurice, her beloved mentor, who urged her to regard imagination as an indispensable tool for the questing detective.

> "We must spark the imagination, Maisie, for the solution to any case is often the one that seems most unlikely… the most amazing facts can be revealed by our fictions, you know. We must be creative in our thinking."[42]

Imagination in detecting is of Aphrodite when, as for Maurice and Maisie, the aim is to restore beauty to daily living for the participants in the case. For Maisie, the case does not finish with the identification of the culprit. Ending a case takes a final "accounting" that goes deeper than even Athena's care for communal values. It includes fostering the ability to recover Aphrodite's spontaneity and beauty-inspired joy, what Maisie now realizes that she longs for herself.

Indeed, Eddie's birth recalls Persephone's abduction into the underworld even more evocatively than that of Christ. While Eddie may well have had qualities of a savior or divine child, his mother, Maudie, was raped as brutally as Zeus forcing himself on some of his victims, or Hades seizing Persephone. It is only Maisie's challenging Maudie to reveal the identity of her attacker that starts to shift some of that persistent trauma. Eddie was bullied and then killed by his half-brother, offspring of a father who violently dominated the streets.

Persephone's descent is invoked several times in *Elegy for Eddie*, because "it doesn't take an earthquake for the ground to break apart and swallow you," for poor people in a London of incredible economic hardship.[43] Maisie meets the bereaved lover of Bart, the murdered journalist, and sees her as falling into an underworld. Fortunately, Eve Butterworth has Artemis support in a sisterhood of single women, and Aphroditean impulses for making her surroundings beautiful. When Maisie last visits her she senses that Eve has made a healing place.

"Persephone" is also the name of Maisie's father's horse, kept under railway arches. Perhaps the presence of this goddess in the novel is to remind us of Aphrodite's dark side in this era of deepening effects of war, past and future. Maisie is herself accused of misusing Aphrodite's gold. Uncomfortable with inherited wealth from Maurice, Maisie is unable to adjust to a privileged lifestyle with James and tries to manipulate her subordinates in efforts to do good.

In particular, she has provided a new house for her shell-shocked assistant, Billy—a gift which backfires when Billy is badly hurt while making inquiries. Without the support of her former neighborhood, Billy's wife falls into an abyss, or depression, a modern take on Persephone's myth.[44] She blames Maisie for Billy's injuries with an anger that is partly the return of a previous breakdown and partly a backlash against having her life transformed by power from above her own station in life.

Maisie does not want to behave like a goddess. She gradually comes to see that she has to be careful not to be seduced by the Aphroditean gold she has used to make people conform to her own idea of what is best for them. This alliance of Aphrodite (seduction of money) and Athena (conforming to norms) is pernicious for Maisie's own wellbeing. However, one aspect of Aphrodite that is generative for her is the goddess's renewal of virginity by bathing.

Several times in *Elegy for Eddie* Maisie has a hot bath in which she reconsiders her case. Here she lets insight arise intuitively, just as the goddess emerged from the sea. Notably, her father-in-detecting, Maurice, is invoked in these baths. In this sense, Maurice's advice is the creative sperm that unites with the sea of Maisie's unconscious to produce Aphroditean knowing. Imaginative, generative knowing emerges as the goddess from the sea of the unconscious.

On the other hand, not all creative imagination in this story is dedicated to Aphrodite as individual erotic connection. Some of it is about psychic change on a collective level. Maisie discovers that Otterburn is secretly preparing England for war with Nazi Germany. He has enlisted writers to change public opinion in a country that is still exhausted from the previous conflict. Otterburn wants Ares's fighting spirit to be conjured from erotic immersion in the beauties of England. He is also building fighter planes, the plans of which were stolen by Eddie and led to his death. James Compton, Maisie's lover, will go to Canada to help test the planes. Lacking Aphrodite's passion for him, Maisie probably won't go with him. In the meantime she comes to feel that everyone is now Eddie, manipulated by powerful forces of which they are only partly conscious.

Ares dominates this era more than his lover, Aphrodite. Insufficiently healed from one war, Maisie seeks Aphrodite's joy for herself, just as she tries to foster it in others. It is time to consider further Aphrodite's imposition of ordeals on Psyche, as that feminine soul wanders the world bereft of her beloved Eros.

Eros and Psyche: the Ordeals of Psyche

Abandoned by Eros for disobeying his prohibition on seeing him in the light (of consciousness), Psyche is forced to realize that she has to serve Aphrodite, who dispatches handmaids Worry and Sadness to her. Then Aphrodite inflicts four ordeals on the suffering woman. Firstly, Psyche is shown a huge mass of mixed seeds and told to sort them in one night. This impossibility is accomplished when ants come to her aid.

Secondly, Psyche has to cross a dangerous river in order to get golden wool from vicious sheep of the Sun. Although tempted to drown herself, a reed gives her instructions on how to complete the task. For the third ordeal, Psyche must journey to the source of two underworld rivers, the Styx and the Cocytus, to obtain their black water. This time Zeus comes to Psyche's aid by sending an eagle. All three adventures offer insight into the travails of the fictional sleuth.

After all, sorting seeds is an image for the discriminating faculty that all detectives require. While every mystery has elements of this task in distinguishing clues from unimportant details, authors as different as Agatha Christie and Lindsey Davis specialize in a plot focusing on a collection of likely suspects. Here, to sort seeds is to discriminate the red herrings from the actual killer. In works such as Christie's *Appointment with Death* (1938)[45] and Davis's not entirely dissimilar *See Delphi and Die* (2005),[46] the intuitive as well as highly rational qualities of psyche are enlisted to weed out the murderer. The ants who aid Psyche are reminiscent of intuition's natural patterning instincts.

A more active and physically taxing mode of detection is modeled by Psyche's second ordeal. Like her, many female authored sleuths face death in their total immersion into a case. They also are sent amongst unfriendly animal, or instinctual, entities in order to find the truth. Anna Pigeon, in Nevada Barr's books, actually does forage in a perilous wilderness in order to preserve human safety and life. In *Winter Study* (2008), she nearly dies in a frozen lake. Later, pursued by a vicious killer, she learns from the sophisticated nature of wolves how to defeat the bestial nature in some human beings.[47]

Similarly, there is a rhythm of physical immersion and suffering in mysteries by authors such as Linda Fairstein, Sara Paretsky, and Marcia Muller. Although all offer urban detectives, Alex Cooper and P.I.s V.I. Warshawski and Sharon McCone are inevitably drawn into corporeal stress and the need to draw on instinctual, embodied, and intuitive knowing, not only to succeed in their quests, but even to survive. Interestingly, the more Hestia oriented Goldy Schulz also sometimes experiences an ordeal of nature, such as hunting for clues in a forest while on the run, in *The Main Corpse* (1996).[48] Here her incarnation as Psyche suffering the wrath of Aphrodite is rapidly assuaged by Hestia when Goldy finds a hut and makes a home-cooked meal for her companions.

Fetching black water from the sources of two rivers of Hell reminds us that the sleuth has to seek out origins, the sources of the crime, and that this will be dangerous and dark. Ultimately, finding the murderer means knowing the source of his or her crime, whether in lust or for riches, or the tangled emotions of families or groups. Maisie Dobbs is a detective particularly devoted to pursuing the source of the problem, and aware that the primal waters of human motivation may be complex, multiple, mysterious, and hard to ascertain.

Like Psyche, Maisie may require Zeus's eagle, with its capacity for more than human vision, to get "above" a situation. In this sense the eagle may be a transcendent insight resulting from a combination of conscious and unconscious powers. As in *Elegy for Eddie* (2012), Maisie uses imagination as her eagle, enabling her to "see" what cannot be rationally known about the overlooked death of Eddie, who was involved with political powers he could not comprehend.

Similarly, in Barbara Hambly's *Fever Season* (1998), set in 1930s New Orleans, Benjamin January becomes sure that the woman he cares for has been raped well before he uncovers confirmation from witnesses.[49] His observing of her behavior and psyche allows his erotically inspired intuition to soar eagle-like into knowing. In Ben's attraction to Rose, here Aphrodite is productively woven into the kinds of ordeals that love brings.

Psyche's fourth and last task for Aphrodite is to visit Persephone in the underworld and procure some of her beauty ointment. In despair, Psyche prepares to throw herself off a tower, when the building itself offers good advice. Psyche must distract Cerberus, the three-headed dog, with honey cakes and take coins for Charon, the boatman, for the trip across the Styx. And at last Psyche gets some help from Eros, who discovers her sleeping after her underworld journey and puts her sleep in a box for Aphrodite. Eros enlists the help of Zeus, who brings Psyche to Olympus, with a drink of Ambrosia in order to make her immortal. She can then marry Eros and remain with him for eternity.

The helpful tower suggests here the way a crime scene can "speak" to an attentive sleuth. So many mysteries begin and continue to invoke matter that "speaks" to them in the discovery of clues. Unfortunately, many mysteries also entail visiting the underworld, which may manifest as depression, well known to detectives as varied as V.I. Warshawski, Maisie Dobbs, and Goldy Schultz. Or when Laurie R. King's Kate Martinelli is banished from the case to which she feels intimately connected, in *Night Work* (2000).[50] Of course the underworld can be literal, as with Goldy Schulz in *The Main Corpse* and Sharon McCone in *Where Echoes Live* (1991).[51] In these works both find murderers in abandoned mines.

Hell can also be other people.
We locked eyes – a split second in hell.[52]

In *Roast Mortem* (2010), Clare Cosi, and her lover, Mike Quinn, find themselves in several kinds of inferno, as Mike's longstanding feud with his cousin and namesake, Michael Quinn, the firefighter, threatens to erupt into a tussle over Clare. Both men are caught in Aphrodite's seductive charms in ways that at first obscure, then finally prove to illuminate, a series of murders by arson.

Even closer to Psyche's encounter with Persephone is Goldy Schultz finding herself involuntarily abandoned by lover and cop, Tom, on their wedding day, in *The Last Suppers* (1994).[53] Here, what should be Aphrodite rooted (and routed) into a marriage explodes into Psyche's ordeal for Goldy, who has to use all her goddess powers, plus her love for Tom, in order to reach that "heavenly" wedding. For this she discovers that an underworld journey is psychologically and physically necessary. Depressed and terrified, Goldy detects that Tom is, in fact, literally tied up underground.

Similarly, time and time again Sharon McCone is forced into the underworld. In *City of Whispers* (2011) she rescues her half-brother from the underworld of drug addicts and crime.[54] *Looking for Yesterday* (2012) finds her lost and depressed when her longtime home in San Francisco burns down.[55] Even more tormenting is the kidnap of associate and friend, Adah, in *Coming Back* (2010). Adah is left vulnerable because Sharon failed to support her, even though her uncharacteristic carelessness was partly due to her own post-injury malaise.[56] In having to develop consciousness through suffering, Sharon is plunged into Psyche's demanding story.

While Adah, locked into small dark spaces, is in a Persephonean role, Sharon has to face the accusations of those who know she neglected her guardian duties.

"I didn't think…I made a bad error of judgment."

It was a difficult admission – like most people, I hate to admit when I'm wrong – but I meant it. I'd screwed up, put my friend and employee in danger, and now she might be lost to me forever.[57]

Here neglect of the duties of love and relationship do recall Psyche's cavalier treatment of Aphrodite that incenses the goddess. And angering a deity has far-reaching consequences. Fortunately, Sharon has evolved a team who know how to set ants to work on mixed seeds, in the remarkable computing skills of nephew, Mick Savage. Sharon herself embarks on physical immersion to discover the sources of this enigmatic crime, in the supposedly dead husband of the woman Adah was investigating. Meanwhile, Craig, Adah's lover, is suffering the anguish of loss and fear for his beloved. Happily, *Coming Back* ends by celebrating Psyche's marriage, the long delayed union of rescued Adah and loyal Craig.

Such an ending recalls how Sharon and Hy's union seeks, not always entirely successfully, to balance Hestia, Artemis, and Aphrodite in one marriage. Yet Sharon and Hy, both pilots, are happiest when soaring in their plane like Zeus's eagle. Indeed, it was in one of these moments that they decided to marry. At the end of *Looking for Yesterday*, Sharon and Hy reach Zeus's eagle's transcendence, in being able to let go of the burned shell of their urban home and their search for a new one somewhere else.

> He turned to me and smiled. "This reminds me of when I changed the plane's course for Reno and we got married."
>
> "It's much the same – a turning point."[58]

With the triumph of the eagle's more than human in-sight, we turn to a ground-based Aphrodite endowed sleuth, in Antonia Fraser's British TV journalist, Jemina Shore.

Case History (24): *Oxford Blood* (1985) by Antonia Fraser[59]

> "She'd made a mistake, if you call love a mistake, twenty years earlier. And that mistake came to cost two lives, three if you count Proffy himself, and it nearly cost the life of another, Saffron. She had to live with the knowledge that her lover had killed her daughter and tried to kill the man she now knew to be her son."
>
> —Antonia Fraser, *Oxford Blood*

Jemima Shore, who prefers to make serious television documentaries under her legend, "Jemima Shore, Investigator," is inveigled into exploring "Golden Lads and Girls" at Oxford after becoming infuriated at the outrageous exploits of a group headed by student, Lord Saffron St. Ives. Unwelcome is the addition of Tiggie (Antigone) Jones, a recent graduate and suspiciously intimate protégé of Jemima's boss, Cy Fredericks. What neither of them know is that Jemima has been the recipient of a dying confession of the former midwife to the St. Ives family, to the effect that spoiled Saffron was in fact an illegal adoption.

Legal or not, adoption is a problem for the St. Ives family, as Jemima explains to lover, Cass Brindley. An adopted child of that era cannot inherit a title or an estate, as Saffron expects to do. Moreover, Jemima feels a responsibility to the dying midwife to do something. Not only is the midwife also a nun who has been denied absolution by an over-zealous priest, but it was a TV program by Jemima on peace of mind before dying that inspired the confession.

Cass groaned. "Oh my God, the ghastly power of television…"[60]

Here Artemis through Jemima and TV exercises her divine power in leading to the mysteries of death. Can Jemima, who combines an Athena social conscience with Aphrodite in her personal life, find a way to bring all three goddesses into harmony?

Saffron proves not to be quite as dislikable as Jemima expects, especially when he appeals to Jemima to find out who is trying to kill him. Likely suspects include those members of his family who would inherit his riches should he die or prove ineligible. These are his father's unpleasant brother, Andrew, a right wing MP, and his very decent son, Jack, a left-winger and also student at Oxford. The situation gets stickier when Saffron announces his engagement to Tiggie Jones, and Jemima is invited to the mansion of Saffron Ivy for a weekend celebration. Tiggie's mother, esteemed classicist Professor Eugenia Jones, is strangely horrified by the engagement.

Still investigating the allegation that Saffron is adopted, Jemima begins a conversation about blood types, only to realize the resemblance between the engaged couple. Saffron and Tiggie must surely be half-brother and sister, making Saffron the son of Eugenia Jones. That evening there is a death. It is not Saffron, but rather Tiggie Jones, who dies of a drug overdose. Of course the circumstances prove suspicious.

Jemima tells scared Saffron that she has to go on with making her ratings-grabbing program for her difficult boss. So she agrees to accompany Saffron to a grand ball as "cover" for her investigation, while privately meaning to look out for him. Typically for an Aphrodite-led woman and sleuth, protecting Saffron includes their making love just before the big event. Aphrodite is here a means to knowing and protecting, since Jemima is present when Eugenia Jones's lover turns up with a lethal syringe to kill Saffron.

> "A long story, a long story from the past. But not, I think, the story you anticipated, Jemima Shore, Investigator..."[61]

Professor Mossbanker, universally known as "Proffy," is a married man with many children. He is also in a long-term relationship with Eugenia Jones, who had in the past refused him as her husband. Father of Tiggie Jones, he is not the father of Saffron. Eugenia once worked briefly with diplomat, Lord St. Ives, who was the love of her life. Saffron is their child. Proffy's murderous spree is fuelled by jealousy and envy of the privileged life at Saffron Ivy. He is consumed by Aphrodite's darkness in sexual possessiveness and desire for the "golden" life.

In turn, Aphrodite has served Jemima better in this quest than her, perhaps always less secure, Hestian qualities. For she admits to Cass that she was misled by sensing tensions at the Saffron Ivy home that caused Andrew Iverstone to be her prime suspect. In fact, the dynamic of envy stemmed from Proffy towards Lord St. Ives. Had Aphrodite not brought Saffron and Jemima together then she might not have realized the truth.

In the end, Jemima also serves Aphrodite by protecting Eugenia Jones from the police and concealing what she knows about Eugenia's children. To Jemima, a mistake in love is all Eugenia seems guilty of, so she reprises some of her divine power in shielding the devastated woman. Saffron, child of illicit love, is, at least, his "official" father's biological son, and his actual parentage remains a secret. "Blood" is satisfied in terms of inheritance, as well as in Aphroditean desires of the body that produced Saffron.

Oxford Blood is a mystery about the mysteriously potent divine powers of Aphrodite to make joy, in the happiness of Saffron's parents in making him, and how he became a much loved son. Yet Aphrodite also can drag one to an underworld of jealousy, abandonment, and even motivate war and less systematic killing. Proffy makes war on the St. Ives family when he realizes that he did not get his Helen of Troy.

Fortunately, Aphrodite offers an instinctual embodied form of knowing that enables Jemima to stop at least one murder.

The final pages show Jemima resisting Aphrodite's darkness, in facing her own anger at Cass's infidelity with Tiggie Jones, notwithstanding her own with Saffron. The goddess has not finished with Jemima Shore, Investigator, nor have we.

Aphrodite and Truth in Mysteries

Aphrodite is the spontaneity of the carnal body and bestows the beauty that inspires desire. What she is not, with her complicated erotic arrangements, is especially truthful. Like Hermes, lying does not trouble Aphrodite, whose energy is as free flowing as water. While not promising eternal love or fidelity, Aphrodite offers those she encounters the choice and challenge of pursuing meaning for them. As Ginette Paris says:

> [Aphrodite]... leaves the responsibility of truth to the one who listens... If we don't accurately perceive, we are guilty of a defective intuition.[62]

While genre insists that truth matters in mysteries, the native devious quality in Aphrodite is arguably also important to the sleuth. Not only do witnesses lie or deceive by omission, so does the detective when undercover. The sleuth meets Aphroditean evasion every day and needs to also embody her wiles. Here Aphrodite offers psychic veils to the sleuth, just as she is habitually depicted with some clothing. Her civilized blend of nature, nudity, with human arts such as beautiful clothes, signals an equally sophisticated attitude to embellishing or artfully adorning the truth.

So Aphrodite offers the pleasures of concealment to the tricky detective as well as the challenge to extend her intuitive Eros into intuiting or imagining what is being withheld. Here the detective's search for truth is erotic, veiled, seductive, and longed for. Yet such is Aphrodite's potent blend of pleasure in veiling her nature that often the truth is desired and simultaneously desired to remain concealed. Not surprisingly then, the longing to know and not to know is often bound up with sexuality and love.

For example, in Carolyn Hart's *The Mint Julep Murder* (1996), a group of suspects in the death of a publisher at a conference all have secrets that give them motives.[63] Yet Annie Darling, gathering her red herrings all in one room like Hercules Poirot, discovers that the husband whose wife is terrified of his discovery of her liaison knows but has refused to acknowledge it. His Aphroditean love demands he veil his eyes from her straying.

Similarly, in *Cream Puff Murder*, Mike, a cop, tries to tell Hannah, his girlfriend and sleuth, that his spending nights in the apartment of nubile victim, Ronni, was entirely innocent.

> "Hannah? You believe me, don't you?"
>
> "Of course," Hannah said, being entirely truthful. She did believe he'd spent several nights at Ronni's place. Perhaps that wasn't precisely what he was asking, but it would have to do for now.[64]

Hannah finds a way to tell the truth and at the same time be evasive on whether she believes Mike's protestations of fidelity. Both are working in Aphrodite to restore their erotic connection that is significantly cemented through detecting.

Perhaps a particularly poignant example of Aphrodite's veiling of truth in the service of detection comes in Jacqueline Winspear's *Pardonable Lies* (2005).[65] Maisie Dobbs has been commissioned by grieving parents to investigate the fate of their airman son in the First World War. In the course of investigating, Maisie discovers that Ralph Lawton was gay in a time and in a family that could not accept it. She finally tracks down Lawton's wrecked plane and meets the badly burned man who had flown it. No longer Ralph Lawton, this man is trying to find peace under a new name. Maisie returns to England to give her report to Lawton's father.

> "And my son was killed."
>
> Maisie paused until Sir Cecil met her eyes with his own. She had considered her words with care. "I can confirm that Ralph Lawton died in the inferno."
>
> Sir Cecil exhaled deeply, though Maisie could see that it was a sigh of relief and not of regret.[66]

Maisie actually offers the facts, but in such a way that Sir Cecil can find the veiled truth he needs. The man could not accept his son's sexuality, and partly for that reason, was relieved to think him dead.

Aphrodite, through Maisie, offers him a crafting of events that enables both men to have some peace.

Therefore, Aphrodite of relationships proves more deeply implicated in women's mysteries than might at first appear. Well known as potentially fatal, this goddess certainly provides corpses in those driven to kill for lust or lucre. On the other hand, I suggest that willful murder is equally an offense against Aphrodite for cutting short the carnal pleasures rightfully dedicated to her. Maybe it is this offense to the goddess that stirs the detectives who enable Eros to guide and inspire their quests. After all, Aphrodite is of the imagination in her ability to see beauty that inspires Eros. Aphrodite's gifts of imagination, her ability to blend art and nature, in crafting naked truth into spontaneous beauty, are gifts that the goddess-inspired detective truly values.

Finally, Aphrodite is necessary to the reviving quality of mysteries; how they overcome the brute fact of death to offer recuperation. Aphrodite is a lover of Ares, not a killer or warrior herself. Although, as with every archetype, she has a shadow, an underworld quality of depression, the pain of abandonment and sexual jealousy, Aphrodite is also the energy of sexual fertility, union, and the body's pleasures as sacred. The canny sleuth needs Aphrodite, not least because she is insufficiently honored in our twenty-first century, metrics-driven world, which is the subject of this book's last chapter.

CHAPTER SEVEN

THE NATURE OF THE 21ST CENTURY
THE SLEUTH AND THE GODDESS AFTER 9/11

C hapter 7 tells a story through eight mystery novels by women, all but one published in the first decades of twenty-first century. Like the masculine hero myth identified by Joseph Campbell, this hero, a woman, is called to survive various conflicts.[1] Unlike most ancient male myths, she does not wholly defeat her antagonist, Death, the ultimate threat to Paradise. However, by voluntarily risking herself, she meets Death and travels through an underworld of suffering that includes her own being. She re-emerges to offer the energy of many goddesses in healing her community of excessive fear of the Other.

Introduction: The Goddesses Return

At the heart of this book is an idea about the re-emergence of the feminine sacred in women's mystery fiction. In this final chapter, I want to look at this rising of the goddesses in relation to the dark pressures of the twenty-first century around two urgent topics: nature and war. Specifically, the matter of war was distilled by the psychological impact of the attack on the World Trade Center in New York, on the 11th of September 2001. Nature, on the other hand, is addressing us in the matter of climate change. While at first glance, the war and environmental crisis appear to be very different topics from what mystery fiction typically provides, both, I suggest, are intimately concerned with the Western ego's relationship to the Other, and to its persistent nomination of the other as feminine.

Mystery novels draw broadly on the rise of the novel form itself in the eighteenth century. Specifically, mysteries stem from the Gothic subgenre with its interrogation of boundaries of all kinds, including that between life and death. In "The Carrier Bag Theory of Fiction," Ursula Le Guin traced the history of the novel as a distinctively feminine form that mixed disparate elements together and forced characters to *relate* to each other.[2] Here the novel digs deep into the roots of the sacred in animism and its multiplicity of spirits in nature as one manifestation of Earth Mother consciousness (see Chapters 1 and 2).

So mystery novels, with their testing (to destruction) of the sleuth as lone hero, contain an intrinsic animistic feminine consciousness of cooperation. Even Sherlock Holmes is not really alone. He has to rely on Dr. Watson and interact with suspects in order to solve a crime. Mystery novels by women, as this present book shows, are hospitable to the various structures of the feminine sacred from a polytheistic era, whose pantheon of goddesses have been long neglected as intrinsic components of the Western psyche.

Earth Mother consciousness of relating to the body as sacred has been persistently marginalized by the dominant monotheistic Sky Father consciousness of separation that inaugurates a foundational dualism. Because Sky Father has dominated, we inherit a culture that regards human nature as discrete from non-human nature. Exploitation of the planet and, arguably, climate change is the consequence. The Western psyche is built on repression of the Other as nature, feminine, other cultures, the body, and matter (mater) itself. This is the dark side of a culture of transcendence that stems from the notion of a single God, one who is transcendent of body and nature itself.

The goddesses emphasized in this book—Hestia, Artemis, Athena, and Aphrodite—are not patterns of transcendence. Rather, they offer a variety of psychic relationships to the body, nature, and to other people. As highly sophisticated psychic structuring, they arrive at a better knowing and orientation to the Other. So, one might expect in women's mysteries to find perspectives and possibilities on nature and others that can offer something new or distinctive. Might women's mysteries be a form of rebirth and renewal for a psyche facing a wasteland of devastated nature and constant wars?

This chapter will investigate relations to nature and responses to terrorism by examining a diverse sample of women's detective fiction in the twenty-first century. We begin with a "cozy" mystery by one of its most reassuring practitioners.

Part One: Home and the Other

In this first part we examine the divine qualities of Hestia in alliance with other goddesses in re-investing nature and human nature with the sacred. A second case history follows in which the trickster genre demonstrates its prophetic potency. A terrorist plot to violate the Manhattan skyline is foiled by a sleuth painfully embodying more than one goddess.

Case History (25): Re-Igniting the Hestian Hearth in *Carrot Cake Murder* (2008) by Joanne Fluke[3]

Carrot Cake Murder tries very hard to provide Hestia, goddess of hearth and home, for a troubled psyche. She arrives in the comfortable shape of Hannah Swenson, owner of The Cookie Jar in the iconically named town of Lake Eden. The story starts in the church, where Hannah and her family have gathered most of the community to hear Rev. Knudson announce his forthcoming marriage to Claire, rumored to be the former mistress of the adulterous Mayor. When it seems as if the couple might back out of confronting the congregation, Hannah gets up and guides the applause for church service into celebrating the engagement.

Here is almost an overdetermined rebuilding of community around the hearth of Hannah's cookies, which she has thoughtfully brought to the church to seal the announcement. The black sheep has been accepted back into the sacred hearth of church and community. Not so fortunate is Gus Klein, long lost brother of the Beeseman clan, now assembling over one hundred relatives for a family reunion. Gus disappeared as a teenager while owing money to his family. He re-appears apparently wealthy and caring little that his parents died still mourning his absence. A great eater of cake, he makes a point of acquiring a carrot cake made by Hannah for the family gathering. Unfortunately she discovers him murdered next to her squashed cake the following day.

"You're going to take the case, aren't you?" says Hannah's sister Michelle.[4] The Beesemans unanimously vote that Hannah detect on their behalf so that they can "get back to normal and enjoy the family reunion."[5] Hannah is the sleuth-goddess who can heal the trauma of murder. She can restore the family, this one so large it becomes a village. Hannah is endowed by the enthusiastic inhabitants of Lake Eden with the ability to restore the hearth of the home. Even cop boyfriend Mike is keen to share information. Although later it seems he is trying to trick Hannah because he knows he cannot stop her detecting, Mike is more properly the masculine principle in Eros mode. He too is gathered to Hestia's hearth, by affection for Hannah and for her comforting food.

In this novel, the nature of Lake Eden is Hestian oriented through Hannah. Its non-human nature is, perhaps not surprisingly, *not* in a binary "other" connection to humans. For example, major characters in Lake Eden mysteries are cats: Hannah's Moishe, and Cuddles, who lives with Hannah's other boyfriend, Norman. During *Carrot Cake Murder,* Moishe provides his own mystery of unusual and noisy behavior. While the otherness of not being certain about the problem is explored, Hannah and Norman decide that the cat is bored by the lack of favorite TV station. A "Kitty-condo" with plenty of toys proves the answer.

On the other hand, the summer weather provides bloodsucking insects that are known as Minnesota's state bird.[6] Hannah discovers Gus Beeseman's corpse because flies are drawn to the spilled blood, although the column of ants are merely interested in the cake. Non-human nature is wild *and* domesticated, incorporated into the family, and proves enthusiastic for Hannah's cooking. Here we remember Hestia is of the earth as home, as well as of the hearth. Hannah adapts to a wild beast living in her apartment; he is not so wild, yet she accommodates his needs that do not tally with hers.

Hannah is also Hestia in refusing marriage to Norman because she is also attracted to dangerous Mike. In fact, in *Carrot Cake Murder,* Hannah's usually dormant Aphrodite aspects, inspired by sleuthing, are also aroused by Norman and Mike. To overcome her shock at finding the body, Norman takes Hannah out on the lake to a beautiful water lily garden. Later, the killer, a Beeseman relative (by marriage), takes his estranged wife to the same place. It was where they got engaged. But this romantic visit has a deadly purpose. Hannah foils what would be the second murder while endangering herself. Thrown into the lake, Hannah hides among the water lilies.

> She floated and her nose came up. She breathed the beautiful slightly sweet smelling air… she was part of a Monet exhibit, and he wouldn't think to look for her here.[7]

Hannah rises from the waters in a beautiful setting like Aphrodite. She is rescued dramatically and erotically by Mike who calls to her "like Marlon Brando yelling 'Stella' in *A Streetcar Named Desire*."[8] Perhaps Hestia alone is not sufficient to recuperate from murderous violence. Both water lily episodes evoke erotic love to counter the offence to Aphrodite of the massacre of the carnal body. On the other hand, *A Streetcar Named Desire* is a work of Aphrodite, who also brings darkness, underworld, and madness. If this is Aphrodite to Hannah, then it's no wonder she resists active sexuality and embraces unmarried Hestia.

Yet in this novel Aphrodite and Hestia both offer a healing incorporation in non-human nature, in the cats as co-makers of the family hearth, and in Aphrodite's liminal waters enabling Hannah to be reborn in Mike's arms, if only temporarily. The "Klein other," in the form of criminal menace, is identified and expelled in what has become in these books *the community ritual* of Hannah catching the killer. Gus Klein actually returned to seize a valuable baseball card. He is killed because he will not pay a debt to a family member. His refusal to honor the hearth and home of Hestia seals his fate in Lake Eden. For in this earthly paradise, sin is either converted, as in Claire moving from mistress to wife, or eliminated, and the trauma of that execution is exorcised on the community's behalf by its centering goddess, Hestia, invoked in Hannah Swensen.

Carrot Cake Murder is indeed a reassuring alternative to twenty-first century horrors focused on nature and the other. However, its stress on mystery as ritual healing of a violated home extends the notion of home. The sacred center shifts from a single goddess with cat, to family, to community, to the planet itself while not entirely ignoring that aspects of the other, as nature, are not human-oriented. This novel suggests that Hestia seeks home in which nature is the stranger at the hearth to be protected.

Case History (26): making the transition to the twenty-first century in *Liberty Falling* (1999) by Nevada Barr[9]

> Under the dispirited flapping of the American flag, she watched the skyline, dominated by the twin towers of the World Trade Center recede, carried away on the wake of *Liberty IV*.
>
> —Nevada Barr, *Liberty Falling*

A terrorist group is plotting to blow up an iconic National Monument in New York, which will change the skyline forever. No, it is not 2001 prior to the felling of the twin towers, on September 11th; rather, it is 1999, and the fictional threat is to the Statue of Liberty, in Nevada Barr's astonishingly prescient, *Liberty Falling*. Here National Park Ranger Anna Pigeon has arrived as Artemis to the big city, because of the threat to the feminine: her beloved sister, Molly, is in a hospital, in critical condition. Anna is staying at the National Park of Liberty and Ellis Islands, as the novel opens with the "unthinkable" possibility of Molly's death. Such an intimate and immediate threat to the feminine is augmented when a small figure falls from the Liberty statue. At first believed to be a boy, it later proves to be a girl named Agnes Tucker. She had been kidnapped from her mother, probably by her father, many years previously. A sensitive park ranger, Hatch, is accused of murdering Agnes. His vulnerability prompts Anna to help find out what really happened.

Ultimately, the threat to the feminine is enacted by a racist and fascist plot to blow up the "Lady," as the statue is known to those who care for her. The gang, headed by Agnes's father, A.J. Tucker, plan to detonate bombs inside Liberty's breasts during a July Fourth celebration of a multicultural America. Their aim is to convert public opinion to their white supremacist obsessions. One of their members is a woman Park Ranger who antagonizes Anna and kills the inquisitive Hatch.

So the threatened feminine here is not in any sense defined as merely women at the mercy of patriarchal men. Rather, this feminine is archetypal and mythical. She inheres in Anna's most intimate and enduring relationship to her usually indomitable psychiatrist sister, in the Statue of Liberty's political resonance, and, importantly, in vulnerability in males as well as females. For Anna is not only engaged by Hatch's deeply troubled psyche over Agnes's death, she also has to recognize that she herself is not Molly's best medicine. For when the doctor tells her that Molly might die because she is losing the will to live, it is a gentle man who proves a most effective nurturing presence.

Anna once had a lover who, on meeting Molly fell in love with her. Molly, like Anna, put their sisterly bond first; both invoked Artemis in their psychic urgency. So Frederick was banished. However, he now re-appears at Molly's bedside, winning her back to life and gradually gaining Anna's acceptance of their love. In fact, Anna, remaining mostly Artemis, in *Liberty Falling* ends up deeply relieved that she can leave the alien environs of the city, because immediate care of Molly can be left to Frederick: "[h]er sister was going to be looked after."[10]

Of course, the feminine who is overwhelmingly here threatened is nature. Yet here nature is Artemis herself as fiercely alive and ultimately victorious.

> Leaves, ocean storms... all the forces of nature – earth, water, land and air – vied for the privilege of being the first to wrest back the island from the clutches of men.[11]

Anna, unable to sleep in Molly's urban condo, finds herself wandering abandoned parts of Ellis Island, finding comfort in nature while reclaiming her own essential natural being. Nature's feminine Artemis includes death, and cannot be defeated. Yet nature's, and human nature's, fertility has amazing powers of rejuvenation, as Molly revives under both Frederick and Anna's attentions. In particular, as Anna learns to let go her feelings of jealousy she is allowing "nature to take its course," in the coming together of her sister and her own ex-lover.

As Artemis, Anna can put her sister first. She can also respond to a "seed" planted in her after she finds a woman barely alive who has scratched a message on her own skin.[12] Ellis and Liberty Islands represent origins to those Americans neither Native nor arriving as slaves.[13] Tucker and his terrorists want to rewrite America's script by destroying her iconic mother who welcomes immigrants. They aim to transform communal guardian, Athena, into dark, faithless Aphrodite, who is allied to Ares in her deadly exploding breasts. Anna faces Tucker: he has guns and a hostage. So the disabled father of murdered Hatch, Jim Hatch Sr., volunteers to sacrifice his life to stop Tucker. Parental love gives energy to Artemis over the terrorists' desire for racism and patriarchy.

Artemis is, as we saw in previous chapters, a goddess of sacrifice, even of human sacrifice. Jim Hatch offers himself as sacrifice, while Anna sacrifices her own possessiveness for Molly's happiness. In preventing the plot, Liberty is saved as statue, for the city and therefore for and by Athena. Significantly, Artemis can here unite with Athena by rising in urban environs to protect and reclaim her wilderness. As sister goddesses, they simultaneously save Liberty/Athena's city.

Liberty Falling asserts that nature is never conquered by human nature; rather, opposition itself structures a war in which nature will reclaim her ground. Not least a factor is that human nature needs to be part of nature, like that of Anna, Molly, and Frederick, in order to live. Those like Tucker, who want to deny the nature of their diverse species, are killers without Artemis's dedication to purity of life. They would deny the feminine and/as Artemis in her innate connection to what is truly natural and wild.

Part Two: Healing New York

Following below are four case histories of novels that I suggest address mythically, and archetypally, and often indirectly, the violence of the twin towers catastrophe. In revealing a psyche-scape of divinities of war and nature, these works battle for human nature and its future.

Case Study (27): the split between North and South, Life and Death in *Three-Day Town* (2011) by Margaret Maron[14]

> She looked like a mummy… a beautiful mummy come
> back to life when he feared she was lost forever.
> —Margaret Maron, *Three-Day Town*

Rebirth and renewal through all four goddesses are the mythical engines of *Three-Day Town*. Here the title refers to the delights of New York as seen by Judge Deborah Knott and Dwight Bryant on their belated January honeymoon. In fact, the book unites sleuths from two series for this author, by bringing Deborah from an extended family in North Carolina to the environs of her distant relation, Sigrid Harald, a New York, detective. The women plan to meet at a party, because Deborah has a mysterious package to deliver from elderly Mrs. Jane Lattimore to Sigrid's mother. However, their initial encounter is complicated by the honeymooners discovering a body in their building.

Three-Day Town displaces the recent history of New York in the trauma of 9/11 (which is never mentioned) in favor of reconciling an older, internal division in the United States between rural South and urban North. Not only does the book follow the perceptions of both detecting figures, but each lends a hand in solving the other's mystery. Deborah, literally living in the crime scene, ends up kidnapped by the killer who is "invisible," or "the least likely suspect."[15] In turn, Sigrid suggests a solution to the problem that reaches Deborah from her Southern family via phone and email. Lee, a young nephew is accused of posting a pornographic picture of a girlfriend on his Facebook page. He denies it, and Deborah believes him because she knows him; he is of her Hestian hearth.

Who had the opportunity to hack into Lee's phone? When Sigrid suggests it could be the girl with the next locker they had all ignored, Deborah realized: "[it] was too logical not to be true."[16] The sleuths connect in this mutual Athena moment. Hitherto they have seemed polarized by Sigrid's reserved Athena and solitary Artemis attributes, while erotic and family-centered Deborah appears possessed by Aphrodite and Hestia. Yet Sigrid, mourning the loss of a lover who has left her unusually wealthy, does betray an Aphrodite moment in her humorous account of laboriously learning to use make up far more recently than Aphroditean Deborah.

Moreover, the case of the murdered building supervisor in New York proves more challenging than it first appears. Indeed, it has more *depth*, since it can only be solved—and Deborah rescued—by going deeper in the apartment building itself. "I didn't go down to the lower level," says Dwight as he proceeds into the underworld to find Deborah wrapped like a mummy in duct tape.[17] Refusing to die, Deborah is reborn from Hades, while the killer confesses that he committed all three murders because people "kept popping up," as if rising from the dead.[18]

Of course Dwight carrying half-conscious Deborah up from the underworld is reminiscent of Eros rescuing Psyche. What is also hidden from the everyday world in this novel is the true nature of the first murder victim, Phil Lundgren. "He" proves to be anatomically female, and yet is accepted as male by his otherwise hysterical and psychologically fragile wife, Denise. Detecting his complete acceptance as male, Sigrid and her team come to see that nature is wilder than human culture often accepts. Arguably, an Artemis insight that serves the cops in *not* getting distracted by Lundgren's unconventionality enables them to see that his integrity, his *wholeness*, is key. Phil, an honorable man, was killed because he wouldn't condone the petty thieving he stumbles upon.

Even more resistant to the detecting process is the far more shadowy matter of Mrs. Lattimore's mysterious package. It contained an obscene, racist bronze statuette made by an immigrant from Nazi Germany who was then in agreement with Nazi hatred of Jews. Later, happily married to a Jewish woman, the sculptor renounced his previous views, and was thought to have melted down all the casts. However Aphrodite had healed the artist while not omitting her destructive side in the surviving sculpture. Offensively pornographic, and with high market value, this dark matter makes a good murder weapon.

On the other hand, the narrative of *Three-Day Town* works hard to weave together Southern and Northern cultural sensitivities. In offering a bed to a partygoer stranded by a snow storm, Deborah and Dwight are praised for their "Southern hospitality" in the middle of Manhattan. Although Sigrid is bemused by the intricate relationships that run Deborah's hometown, she does begin to use these connections when suspecting a mystery about the health of her grandmother, Mrs. Lattimore.

As kin, and as Hestia of hearth and family, Deborah is prepared to tell Sigrid the truth: that Mrs. Lattimore is dying of cancer and concealing it from family for fear of being forced into unhelpful treatment. As Athena, Sigrid knows death as necessity, while as Artemis she understands the need to make one's own way into that particular underworld. She tells her team that not all Southerners are racist. She and Deborah begin to trust one another enough to allow the goddesses they constellate to work together creatively.

In *Three-Day Town*, the only "firestorm of explosions" is Deborah and Dwight making love.[19] Snow makes Manhattan look beautiful and pristine.

> ... Sigrid... found herself caught up in the beauty of bare tree limbs etched in white against the dark brick or stone of the buildings.[20]

This note of "limbs" emphasizes the Aphrodite cloak New York wears that entices even austere Sigrid. Finally, the mysteries solved, Deborah and Dwight decide to cut their week short, and return, as Deborah puts it, to "our son," Cal.[21] They leave from New Jersey, which gives them a view of Manhattan that appears purged of pain even in its diminished towers. Here, again in the final pages, nature is more creative than the nature/culture binary that Phil Lundgren transgressed.

For Dwight reminds Deborah that she referred to Cal as "our son," when in fact he is her stepson. Deborah responds that he feels like her child, although she will be careful of his need to remember his biological mother. So Dwight reveals that in his long previously hopeless love for Deborah he used to pretend that she was his first wife. Cal *is* Deborah's son from her expansion of psyche into the maternal through her Aphroditean relationship with Dwight and her Hestian talent for centering home and hearth. He is also her son through Dwight's Aphroditean passion for Deborah when he was married to another.

Sigrid, on the one hand, finds Hestia trying. On the other hand, her growing tolerance for her new roommate as participant in her *home* helps her see the possibility of the invisible man suspect in time to prevent further tragedy. So together, Deborah and Sigrid bring four goddesses into being and knowing in ways that enact a model of healing. For traumatized New York these detectives revision deep cultural divides: those between North and South, and also the rift between nature and culture, in Phil Lundgren and Cal. By digging . deep into the depths of the apartment building, the sleuths discover the underworld of petty crime and educational deprivation.

From this darkness, the mummy reveals a living woman; a rebirth of goddesses to renew the split social, historical, familial, and psychological worlds of the twenty-first century.

Case History (28): the trauma of survivors in *Roast Mortem* (2010) by Cleo Coyle[22]

> [Michael Quinn] lost every member of his firefighters company when the first tower fell.
>
> —Cleo Coyle, *Roast Mortem*

Roast Mortem addresses head on the trauma of post 9/11 New York firefighters in the character of Captain Michael Quinn standing for the sufferers. He encounters Aphrodite, and in particular her erotic triangle with husband Hephaistos, the blacksmith god, and her lover Ares, the warrior. The "fire triangle," as Captain Quinn tells Clare Cosi after she rescues herself and her barista from a burning coffee shop (not her own), is oxygen, fuel, and heat. Yet after the captain and his namesake and cousin, Mike Quinn, the cop, have come to blows over Clare, she is told: "*you* my friend are in a fire triangle" (italics original).[23]

What is triangular in Clare's relations with the two men with identical names is that both seem convinced that she is Aphrodite, capriciously capable of sharing her affections. Firefighter Michael is convinced of her erotic availability because, in his trauma, he cannot relate to women any other way. And Mike, the cop, finds it hard to trust her because he has been betrayed, notably by his wife sleeping with his fiery cousin in the past. Hephaistos also haunts the tale in that faulty fire equipment proves deadly, another clue that Aphrodite's love triangle is constellated. Moreover, dark Aphrodite drives crime, by the seduction of money, lack of fidelity, using sex to try to appease pain...when arsonists appear to target coffee shops.

Fire to Captain Michael is a "she beast," by nature feminine and lethal.[24] *Roast Mortem* is a novel in which fire is also life energy and sexual passion.

> And that's when it hit me just how much Mike Quinn [the cop] enjoyed igniting blazes in my hearth.[25]

Mike Quinn, the cop, comes from a family of firefighters, so why didn't he follow his father and cousins into their profession? Unravelling the taut threads between the Mike she loves and his "evil twin" cousin, Clare needs also to be Athena, who can re-weave a community and family tragically split over competing loyalties.[26] In *Roast Mortem*, both Mike and Clare share the motif of questing knights attempting to right wrongs, although in Clare's case she refers to Cervantes mock-heroic, Don Quixote.[27] After taking the crime aspect of firefighting into the police, Mike really incurred Michael's wrath when refusing to cover up the dangerous drunk driving of Michael's younger brother, Kevin. Enmity is then compounded by Michael taking an Ares role to Mike's Hephaistos with the latter's former wife.

The novel turns on interrogating the uncontrollable nature of fire as feminine and as sexuality. Firefighters invoke Hephaistos in their reliance upon tools to tame the she beast, one of which Captain Michael likens to King Arthur's Excalibur.[28] Yet when the equipment manufacturer starts saving money by using substandard materials, a fireman dies and another is killed to cover up the plot. With a killer driven by the lust for money and sexual obsession for an equally avaricious woman, dark Aphrodite and Hephaistos betrayed and betraying (his skills with tools) stalk Manhattan.

Fortunately, Clare as honorable if slightly comic questing knight is Aphrodite blended nobly with Hestia and Athena. She is able to break the fire triangle with Michael and Mike because she refuses Aphrodite as her only goddess, and so prioritizes fidelity to Mike. Indeed, managing a coffee shop is more than a job to her. It is a vocation standing for the Hestian qualities of home, center, and hearth, or as she puts it, "constancy, dependability—maybe even loyalty."[29] Therefore, by breaking out of the dark Aphrodite cast on her by two wounded men, Michael by 9/11, Mike by his faithless wife, Clare also unpicks their "evil twin" chains.

In fact, Clare has to sort out the problem of the two men as "mirror images… neither able to change the other," because after Mike punches Michael for groping Clare, she discovers Michael beaten nearly to death, for which Mike is arrested.[30] Moreover, after several coffee shops get torched, and a letter claims the arson to be an ideological and terrorist campaign, Homeland Security targets the radical lover of one of Clare's favorite customers.

Clare is certain that the outbreak of coffee shop burnings is not a continuation of 9/11 type terrorism. She, in Athena's rationality, determines that trauma such as 9/11 has no place in her city now. Indeed, she and her friends specifically pivot the crime plot away from that kind of urban trauma to domestic (hearth) fires/passions out of control. After all, if Hestia and Hermes are not holding their polarized balance, then tricks with fire, breaking the safe circle of the hearth and burning homes, are all too likely. Finally Clare realizes that she is not alone in a fire triangle. One of the men wanting insurance money and the daughter of a coffee shop owner have indeed played a Hermetic trick, in swapping arson targets with another owner.

At last only the collision of Michael and Mike needs to be disentangled by Clare. As Athena she notices that Hephaistos's once good work on fire equipment has been falsified by a dark Aphrodite's lust for more profit. *Roast Mortem* invokes and then disposes of the fear of another 9/11. Fire bombs begin and end *Roast Mortem.* However, the second fire bomb is prevented from exploding in Clare's coffee house by her ex-husband Matt. Despite being a man too consumed by Aphrodite for fidelity, he nevertheless manages to protect his hearth by tossing the smoking bomb into an abandoned building. If not loyal in sexuality, Matt is learning Hestian qualities from the coffee shop he and Clare will jointly inherit.

Telling Captain Michael that he made Mike into the enemy, when it was his own fiery nature that terrified his brother into messing up his trajectory into the fire service, Clare separates the two men from casting each other into a diabolical shadow. At the end of *Roast Mortem* Clare and Mike affirm their fiery natures in the form of the Hestian hearth: they go home.

Nature as feminine and fire is uncontrollable; death is a necessity, as Athena knows. But in a world of many goddesses and gods, divine qualities animate psyche and matter. Healing is to charm the goddesses, so that the Aphrodite, so abundant in the fiery love-making between Clare and Mike, is tempered with their possession of, and by, a Hestian hearth.

Captain Michael renounces his misplaced hatred and moves out of firefighting to where he cannot lose any more men. Settling near his brother, the wounded Ares/warrior of New York is healing through giving Hestia another chance.

Case History (29): Aphrodite's Comic Resolution in *Fearless Fourteen* (2008) by Janet Evanovich[31]

> "We're not playing *Minionfire* anymore," Zook said. We're in charge of homegrown security now. We got weapons to make and posts to man. We're keeping the integrity of the crime scene. We're protecting the house."
>
> —Janet Evanovich, *Fearless Fourteen*

In the Burg of Trenton, New Jersey, Aphrodite and Hestia are invoked in a world where family values include the family of the Mob and everyone is related to someone who fits "people into cement overcoats."[32] It is a comic universe in which trickster values of survival rule. Here police and civilians alike adopt trickster attitudes to the law that enable mobsters, their relatives, and imperfect human nature to rub along. So when a rumor spreads that cop, Joe Morelli, literally has the keys to a nine million dollar bank heist buried on his property, he ought not to be too surprised to find his girlfriend Stephanie's Grandma Mazur among those digging up his garden.

Stephanie Plum, a not very competent bounty hunter, is, like Aphrodite, poised between two lovers. Morelli, once Aphrodite-like himself and faithless, is now sprouting Hestian qualities and keen to offer Stephanie his home. He even accidently proposes marriage to her. Morelli is a more devastatingly attractive Hephaistos to Ranger's (her other lover) warrior Ares. Not bound by any rules such as those to which cops adhere, Ranger, "exceptionally intuitive and doggedly aggressive," combines Ares's expertise fighting with Dionysian lack of restraint and magical powers.[33]

Both men desire Stephanie. However, in *Fearless Fourteen* the overt site of contention is Morelli's home, which develops more Hestian sacred qualities the more it is threatened. First of all, Morelli and Stephanie have reluctantly taken in fourteen-year-old Mario, known as Zook, when his mother cannot make bail. Eventually her crazy brother, Dom, who is convinced that Morelli is Zook's father and that Morelli's house is rightfully his, gets Loretta released. Only then she gets kidnapped by Dom's partners in a former bank raid who are looking for the money.

Untrustingly, the partners hid the money so that no one of them could get it without the others. Yet one of them, whose identity is unknown, starts killing them off. Indeed, Morelli's house proves, like the Burg itself, to harbor a deadly underworld. Stephanie finds a dead body in the basement. She is accompanied by Moon, an old friend who has become Moondog, the Griefer, in Zook's *Minionfire* online game universe. Once arch enemies, Moondog is finally routed by Scorch, the gaming name of Grandma Mazur. She was introduced to *Minionfire* by Zook.

> Grandma... [b]lack jeans, black boots, black T-shirt with WARRIOR written in gold and red flames across her chest... She looked like the Grandma from Hell.[34]

Yet action moves from the relatively stable online cosmos to center on Morelli's house, where Zook, Moon, and Gary the Stalker unite to form Homegrown Security (their response to external threats is, of course, reminiscent of Homeland Security). Wisely denied actual weapons, Homegrown Security protects "the border" with a potato rocket made by Moon. In 2008, *Fearless Fourteen* re-stages the twin towers assault in mock heroic bathos as a battle for illegal loot under a small suburban home. Compounding the theme of expelling trauma is Lula, ex-ho' and companion of Stephanie, whose experience of dark Aphrodite has left her struggling to find the goddess's joyous aspects.

Lula's lust for a life of indulgence in food and sex is linked to her traumatized past, which included a brutal beating and rape. She reaches for Athena but cannot link the communal deity's rationality to the centering and grounding of Hestia. Instead Lula finds herself caught up in an illogical momentum, seemingly unable to stop a sequence of events aimed at marrying her lover, Ranger's subordinate, Tank. Lula reveals the psychic pain of an urban world existing with an underworld of drugs, crime, and prostitution. She is a survivor needing to trick herself into a rationality barely covering her vulnerability. So she finds herself committing to a big wedding when, in truth, she does not know if she wants to marry. Fortunately she still sends Stephanie for Cluck-in-a-Bucket fried chicken, because everyone knows you don't gain weight on a Sunday.[35]

It takes a more decisive Athena and, surprisingly, Artemis, to put back Aphrodite and Hestia into their uneasy peace with the Burg's "family values," and metonymically, for Stephanie, Morelli, and Ranger. While Stephanie cannot be described as a dedicated and thoughtful sleuth, when Loretta's toe is sent to Morelli's house with a demand for the money, she decides that a line has been crossed.

> "Good pedicure," Lula said...
>
> "I don't want Zook to see this," I told her. "He's just a kid. He doesn't need this. And I can't stand around and let them chop off Loretta's body parts. We have to find either Loretta or the money."[36]

Partly motivated by a Hestian familial feeling for Zook, Stephanie starts to use her head. Her Athena qualities of reason and confronting the Furies that are cutting up living bodies lead her first to find the camera trained on Morelli's house, secondly, to spotting the killer. It's an inside job, with the crime scene tech deciding to subvert Burg ethics and thus open the doors to the criminal underworld. And when he escapes with the loot, it is Moon with his potato gun who smashes the windscreen, causing the getaway van to crash into a Deli and nine million dollars, along with frozen pizza, erupt into the sky. One pizza wallops Brenda, singer turned Reality TV star, so fulfilling the "beware a flying pizza" prophecy of stalker, Gary, he who developed his powers after being struck by (Zeus's) lightning.

Moon is an unlikely Artemis. Yet, in *Fearless Fourteen* he is a decisive and effective hunter. Stephanie is right when she says that his strangeness is a way of keeping his virginity of psychic integrity, "protecting his oneness."[37] His name, Moon, of course, is a clue to his divine allegiance. No one really cares about recovering the money, so Moon, who lives by his own or nature's law, grabs some of it and gives enough to Zook and Loretta to make their lives easier. He also gives a lot to a charity that spays cats. His affinity with non-human nature is indeed mysterious!

Loretta, rescued, proves to have all her toes. (Where the killer got the toe he sent remains a mystery.) *Fearless Fourteen* is about the restoration of a violated home, be that Morelli's or the Burg itself. Hestia rubs along with Aphrodite as sexual desire both reinforces and threatens families. When the liminal divisions between suburban society and criminal underworld become unstable, Artemis and Athena are invoked. *Fearless Fourteen* addresses the trauma of twenty-first century terrorism in a combination of comic deflation and satire, such as the potato slogging Homegrown Security. It also reveals the trauma in characters such as abused Lula, kidnapped Loretta, scared Zook, and mentally ill Dom.

What remains suggestive is that the real threat comes from within. Whether it is Aphrodite challenging the potential of the Hestian hearth for Stephanie and Morelli over her desire for Ranger, or the corruption within the police, it is the psychic borders of the underworld that the mystery needs Athena to re-weave.

Case History (30): Ghosts of the East in *Death Dance* (2006) by Linda Fairstein[38]

> "[Mike] doesn't have a better friend than you. We got
> to think for him now, we got to be there when and if
> the center doesn't hold."
>
> —Linda Fairstein, *Death Dance*

In Linda Fairsten's *Death Dance* the center is not holding. Not only is Mike, the central figure in the detecting triumvirate (of New York District Attorney Alex Cooper, and cops Mercer and Mike), still stuck in grief for a dead lover, but there has been a murder at the Metropolitan Opera. A troublesome ballet dancer is found hacked into pieces. When a second crime scene arises at the City Center of Music and Drama, Alex ends up in its iconic dome as one of two hostages of an unstable pair of killers.

These centers may not hold because they are haunted by the past. "Every theater has a ghost," we are told when it seems that theater owner, Joe Berk, and his acrimonious family are involved in nefarious crimes.[39] The Met has a ghost of an oppressed mistress. Now Natalya Galinova has been thrown into an air conditioning shaft, so blocking the breath of the huge building. After Berk apparently "dies" of electrocution, all are unsurprisingly shocked at his revival. "You look like you've seen a ghost," he says.[40]

Alex Cooper, prosecutor of sex crimes, combines formidable powers of Athena and Artemis. She puts her intense professional intimacy with Mike above the possibility of erotic attraction. In fact she sacrifices that chance to her devotion to her job for the city (Athena) and for the mainly women victims (Artemis).[41] Unlike Aphrodite, Alex is satisfied with drab surroundings in her office provided by the city. She relies upon rational argument in pursuing her cases in court. Her preference is for Athenian values, in order to counter the Furies inherent in rapes and sexually motivated crimes. This preference is accompanied by Artemis's loyalty to women pursued for their bodies, and by the uncanny way death follows.

> "Don't you ever feel spooked by this?" she asked me.
>
> "By what?"

"By death, Alex. How death seems to follow you wherever you go."[42]

In *Death Dance,* the dead and ghostly cluster around Alex, signaling a nature that cannot be denied, even in New York. Alex is also Artemis who guides us through the mysteries of death. In this novel of the center in peril, can Alex as sleuth, so dominated by Athena and Artemis, hold the psyche of the city? After the terrorist attack in 2001, can Hestia holding the center be reignited in the haunted theaters of a traumatized city?

At one moment, as if to emphasize the extreme fragility of the Hestian hearth, Alex is forced to leave her apartment at night because of an electrical fault in the elevators, later found to be criminality targeted at her. Rules changed subsequent to the 9/11 attacks actually ensure that the relative safety of a locked apartment must be violated.[43] Yet *Death Dance* nevertheless possesses a central Hestian drive of countering the wounds haunting this particular city. For while Alex consciously represses her need for a personal home (and chooses lovers unlikely to supply one), Fairstein's novels instead offer police officer Mike, a devoted historian of New York and Military History.

In book after book Mike is the Hestian spider providing historical webs of narrative to bind past and future in a way that creates a holding warmth in Alex's mainly Athenian weaving of the case. Here Mike begins *Death Dance* himself haunted and less able to offer his historical imagination of what is not seen: the underworld beneath the streets and the past that built it. However he is still able to suggest how a more acute historical sense might have helped on 9/11.

> "And in World War Two, the army mounted antiaircraft guns on top of High Mountain in case the Nazis made it over the ocean. They should have kept the frigging things there to welcome those Al Qaeda bastards in 2001. A lot of people I care about might still be alive."[44]

Here a historical sense is a protective sense, a Hestian guarding of the hearth or *center.* The deep drive in *Death Dance* is to the center in order to save it. This trio of sleuths needs to restore a sacred hearth for the city. Alex's commitment to Athena is directed to protecting the city from (sexual) Furies; her Artemis is in protecting women and relating as sister goddess to friends. She needs Mike, including her barely concealed erotic Aphrodite attraction, to offer a rational way of re-forming the Furies of chaos into a Hestian sense of New York as home. For crucial to solving the trauma of murder at the cultural centers is a bizarrely theatrical adoption of eastern mystery. This time Mike is student, rather than teacher, of the story of how a huge Islamic Mosque became the City Center of Music and Drama.

In nineteenth century America, a seventh century "Order of the Mystic Shrine" was revived as an offshoot of the Freemasons. Building mosques and setting up rituals, they constructed Mecca Temple on 55th St., later to be renamed the City Center. It is here, in its massive walled off dome, that Alex is held hostage. So Alex finds herself a prisoner in an Islamic building full of the trappings of arcane rituals, not to mention ghosts. In New York, the past haunts the present. This past includes the trauma of 9/11, metonymically symbolized by Mike's grief for his lover, Mike, who lost people he "cared for" that momentous day. The past haunts the present in the fabric of the city, its legacy of immigrant communities and the imagination of the other they engender.

What may keep these painful ghosts from fragmenting the city is a way of *knowing* or psychically integrating the past-haunted psyche via its population of goddesses and gods. Alex, like Artemis in Iphigenia, finds herself in danger of being a human sacrifice to the hunt. Fortunately, her devoted "brothers," Mike and Mercer, rescue her by working out the precise historical materialization of the past in the much re-worked historic dome.

The Center and the "center" are cleansed of contamination by those whose avarice destroys families and cities. Islamic motifs remain at the "Center" of New York. Perhaps the tribute to Islamic culture is *Death Dance*'s attempt to make a Hestian thread of connection to combat the message of the violent opposition acted out on 9/11.

At any rate, *Death Dance* insists on Athena's re-weaving and consequent conversion of the Furies. It does so by Alex's sleuthing as Athena by design, her conscious role, and Artemis by default, her invocation of nature, sister, connectivity, and death. She reserves her Aphrodite senses for her Martha's Vineyard cottage, whose beauty is offered as a place of healing and peace.[45] Here she will marry her friend, not in the sense of getting married, but by officiating at her (sister in Artemis's) wedding.

Although at the end of the novel Mike remains unhealed of his ghosts, his friendship with Alex and Mercer is restored, and so his center will hold.[46] The Hestia they evoke for each other in their context of work and the city has been enough to revive the City Center or, arguably, to model stopping the breakdown imaged in 9/11, when the iconic center of the Twin Towers was destroyed.

Yet engendering some of Hestia's sacred hearth powers for New York is not enough to still all its unhappy ghosts. Mike, city historian and dedicated cop, offers the last few words. He is back to work, reconnected to life, but haunted.

> "Too many ghosts in those theaters – way too many. And I haven't even learned how to deal with my own."[47]

Part Three: Nature and War

In two final case histories, we look at how the genre maps the complex changes in attitudes toward nature and war. Tracing an attempt to find a way through the major challenges of our time, how might mysteries invoke the goddesses?

Case History (31): Broken Promise Land in *Breakdown* (2012) by Sara Paretsky[48]

> "Most people in this country are decent and don't act on hate."
>
> I thought of lynchings, and the murders of abortion providers, and the assaults on Muslims and gays, but I was too tired to argue. I needed what was left of my wits to try to understand what he really didn't want to come to the surface about Wuchnik's death.
>
> —Sara Paretsky, *Breakdown*

"In their death they were not divided," is a biblical quotation used as epigraph to George Eliot's nineteenth century novel, *The Mill on the Floss* about a brother and sister who are finally drowned in each other's arms, reconciled after years of anger.[49] In Sarah Paretsky's V.I. Warshawski mystery, *Breakdown*, a divided country is focused through three pairs of siblings. One of each pair is murdered. At the heart of the complicated story of historical, political, and psychological breakdown is the thread that unites all three apparently unconnected cases.

In a scrap of paper left by the first victim, V. I. finds the quotation that she later learns comes from another victim; it will prove a clue to the death of a third. "Not divided," haunts the novel as a bitter irony, given the profound hatred that spans the pogrom of Jews in Lithuania during World War Two, a right wing polemicist inciting resentment of immigrants and the spread of lies by modern technology. So rapid, pervasive, and persuasive is Twitter and the Web that fact and paranoid fantasies are almost impossible to separate.

Now fifty, V.I. is tired throughout *Breakdown*. Yet even now she does not ignore appeals for help, such as from young cousin Petra, concerning twelve-year-old girls in her book club who devour stories about vampires and then head for a lonely graveyard to perform a ritual. Another frantic call comes from a mentally disturbed friend, Leydon Ashford, who summons her in the name of Artemis the Huntress (V.I.'s middle name is Iphigenia, an avatar of the goddess).[50] Discovering the girls, V.I. also finds a dead body murdered with a stake in the heart.[51] Victim Miles Wuchnik, a shady Private Investigator, will later prove to have had a fateful encounter with Leydon when she was incarcerated in Ruhetal, the state mental hospital.

Meanwhile, the young vampire fans include the granddaughter of holocaust survivor, Chaim Salanta, whose fortune is funding the Senate campaign of the mother of another of the girls. The politician, Sophy Durango, an African American, and Salanta are subjected to a hate campaign on TV by Wade Lawlor, who soon includes V.I. in his vitriol. Lawlor's show is part of a right wing media empire threatening the career of a longtime friend of V.I., journalist Murray Ryerson. He, in turn, has been refused permission to investigate cases at Ruhetal of inmates serving indefinite detention for crimes for which they have not been convicted. For the mentally ill, justice is halted when they are declared incapable to stand trial. It is V.I. who discovers that one of these cases is that of the impaired man who was universally agreed to be responsible for the death of Wade Lawlor's sister.

For most of *Breakdown*, the discouraged V.I. does not feel like any kind of goddess. Invoked as Artemis by Leydon, the exhausted sleuth finds her friend nearly dead in a college chapel after being pushed from a balcony.[52] In fact, V.I. slowly learns that she cannot succeed in this case without at least two goddesses: Athena and Artemis. Minus Athena, V.I. cannot control her rage that she recalls her father telling her makes for a bad cop.[53] Rage only fuels the Furies. And without Artemis, V.I. will fail as a hunter. She remains passive before terrifying events, as she tells a new confidante, Dean Knaub, whom she meets after Leydon's almost lifeless body is taken away.[54]

Later, V.I. tells the Dean of moments of mental dissociation, of being in another place rather than in immediate time.[55] The Dean calls it advanced spiritual practice, a clue to the dark-edged nature of the goddesses in *Breakdown*. For here Artemis is divided between lunatic Leydon, she of the deranged moon, and V.I. herself as hunter, gradually re-learning to track a literally treacherous wilderness of human nature. It is only when Artemis is re-united, when V.I. realizes that Leydon's apparent paranoia about listening devices is actually divine *divination*, or intuitive sleuthing, that progress can be made connecting Wuchnik's murder to Leydon's.

Of course in a novel about bloody rituals in moonlight and killings that look like human sacrifices, Artemis is not confined to the sleuth as huntress. Lunacy, technology, and magic are linked via the young girls whose phones are hacked, even as they hope a shapeshifter called Carmilla—a figure taken from novels—can save them. They fear an America that hates Jewish Holocaust survivors, as well as recent, possibly illegal, immigrants from Poland. Artemis in *Breakdown* is feminine energy of magic, dreams, intuition, even lunacy that dissolves the boundaries between nature, human nature, and technology. Artemis is dark here in madness and human sacrifice, as V.I.'s prescient dream of Leydon indicates.

> In their midst, I discovered Leydon, hanging crucified next to the social worker's great-grandmother. Fire was bursting from her hands and feet.[56]

Yet V.I. has to be Athena as well as Artemis if she is to tackle the Furies that offer only endless violence and who have a claw in her own traumatized psyche. Ultimately, it is Athena whose reason can pierce the prejudice that is keeping Tommy Glover in Ruhetal for the murder of Wade Lawlor's sister, Magda. Prejudice blinds—so says the retired fireman who ought to have realized that Tommy was having his photograph taken with his beloved friend, Good Dog Trey, the fireman's mascot, at the time Magda was strangled.[57] So-called "good people" in *Breakdown* prove not to know the depth of prejudice in themselves, not to know their own underworld. Even V.I. assumes Leydon's ranting is meaningless lunacy, not the divine frenzy it proves to be.

Breakdown shows that human sacrifices occur when we are ignorant of the underworld of psyche and society. Chaim Salanta's granddaughter, Arielle, is tricked by unscrupulous (trickster) detective, Wuchnik, and eventually kidnapped, all because Salanta wants to keep her safe, meaning ignorant, of his darkness and that of her historical origins. Persecuted by Lithuanians and Nazis alike, his mother sacrificed herself to a brutal member of Lithuanian death squads in the hope of saving her children, one of whom, Chaim, became the same man's sex slave until he could escape. Years later in America, Tommy Glover was the innocent sacrifice of a community who preferred not to know that a brother could kill his sister out of sexual jealousy.

Of the three brother-sister pairs, Leydon's brother persecutes her for her unrespectable Artemis indifference to his desires for status; Miles Wuchnik loves his sister, yet has colluded with her sacrifice of her needs to caring for aged parents; and Wade Lawlor killed his beautiful sister for preferring a lover over him. "In death they were not divided" is Leydon's reference to Lawlor keeping his sister by murdering her, perhaps a metonymic comment on the dire state of the feminized other of modern America.

Breakdown ends with Leydon's funeral and Knaub, something of an emerging positive masculine figure for V.I., saying that the love that binds people in their imperfection can span life and death. So in the sense of enduring love between dead and living, in the final words, "truly, in death, we are not divided."[58]

What the goddesses do in *Breakdown* is overcome the divisions and barriers that we humans erect between sane and insane, feminine and masculine, human and non-human, magic and technology, conscious and unconscious, life and death. All four goddesses discussed in this book play a role in offering a solution to the divisions inscribed in this novel. Artemis and Athena are V.I.'s chief inspirations. Once V.I. is one with Leydon's in-sight, she can use her head (as her father advised) as Athena and weave threads that can restore the ability of this community to continue. Hestia not only is the spider-like web that V.I. must detect as linking all diverse participants, she is also the sense of home that is crucial to getting Chaim to talk, to getting the girls to reveal their secrets, and to the final group of conspirators addressing the problem of murdered Aphrodite/Magda.

For dead, beautiful, erotic Magda was found by Tommy in the lake. She did not rise from it living. And for most of *Breakdown* V.I. is separated from her current lover, Jake, haunted by the fear of eventually losing him to a younger woman. V.I. bravely refuses to resent the young in her Artemisian rescuing of the young girls and then in tracking down Arielle. She too is thrown into the same lake as Magda and left for dead. But thanks to her friends and a disgruntled fisherman, rumors of her death prove premature. In a superb performance of trickster triumph, Murray gets a script about the life and death of V.I. approved, then goes off message.

When V.I. walks on stage aiming to surprise Wade Lawlor into admitting guilt, Aphrodite, in her trickiness over truth telling, is reborn as sleuth. Only partially successful, V.I. has to let the police take over securing evidence for his multiple killings. Renewed from water, like Aphrodite, V.I. is also able facilitate the rebirth of Tommy as innocent victim, not just one more human sacrifice to our divisions.

Ultimately, *Breakdown* shows that in a brutally divided twenty-first century nation cut off from nature both human and non-human, the gods are also dark. On the other hand, this mystery is also a cradle of goddesses whose divine powers can restore, renew, and bring rebirth. Rising from the lake to a grudging male, then the tender arms of devoted friend Lotty, V.I.'s mystery constellates Aphrodite momentarily helping Hestia. They are supported by the courage of Artemis (Leydon's lunacy making V.I.'s hunting divine) and Athena's reasoning (refusing to be defeated by the Furies' rage) that got V.I. into the waters of her unconsciousness.

In *Breakdown*, the lunatic and the naïve girls are correct that vampires are real. They have new ways of biting, via Twitter and Facebook; they also survive decades underground in hatreds from another time and place. Fortunately, just as real are the goddesses and the mysteries that invoke them.

Now for one last case—a mystery that seeds hope while being set in a time of approaching darkness.

Case History (32): Revisioning the Other, in *Leaving Everything Most Loved* (2013) by Jacqueline Winspear[59]

> "[A]t the end of the day, I do my job so that people like Usha Pramal have a voice. I cannot bring back their beauty to this world – and even the most ragged soul was once beautiful, Mr Ashley. But I can stand up and find the truth. Sometimes I'm successful and sometimes not. But I do my best."
>
> —Jacqueline Winspear, *Leaving Everything Most Loved*

Maisie Dobbs, psychologist and private investigator in 1930s London, is successful in finding who shot Usha Pramal and left her body in the canal. Usha, whose name is that of a dawn goddess, is repeatedly called "daughter of heaven." She is a highly educated Indian woman, brought to London as a governess, but who ends up in a hostel for immigrant women who've been discarded by their employers. With a healing touch and "imagination without limits,"[60] her refusal to hide her own power to inspire love and connection of all kinds results in her death. Maisie is told that the likely motive was Usha's goddess-like potency that provoked love, one that proved unendurable to a lover or to one close to him.[61]

Another Indian woman suggests that the killer cannot bear the shadow within, revealed by the mere presence of Usha. It proves that the long dark claw of World War One reaches even to a woman who lived far away from where it occurred. Demons possess war veteran Arthur Payton, who in India beat his wife and tortured his small son. After Payton's suicide, his widow married Jesmond Martin. This man was previously rejected by Usha, not because she did not love him, but because he failed to understand and respect her family's customs.

Usha travels to England as governess to fulfill a dream of earning enough money to start a school for girls. Meeting now married Jesmond by chance, she leaves her position rather than be close to him. Yet her good intentions are betrayed by the complicity of a so-called clergyman, friend, and dependent upon Martin's money and silence. For a price he manipulates Usha into the Martin household. It is Martin's adopted son, Robert Payton, abused by his father's war trauma, who cannot bear Usha's and Martin's rekindled love. The devouring psychic legacy of war, those memories termed dragons by Maisie and her friend,

Priscilla, extinguish the luminosity of Usha and her devoted friend, Maya Patel, also shot dead by the boy.

Strongly and fatally present in *Leaving Everything Most Loved* is Aphrodite, she who brings the beauty that inspires love and desire. Indeed, Usha dies because she both is and is not Aphrodite. She has the goddess's radiant beauty, without her immortal stamina. Unlike Aphrodite, Usha does not survive her entanglement with Ares, the god of war. For, as Ginette Paris pointed out, modern warfare does not sufficiently constellate the warrior lover of Aphrodite. Instead the poison gas of World War One and the disembodied technology of today offer something more infernal, because they are less integrated with Aphrodite's healing sexuality.[62]

More successfully, Aphrodite here is Maisie in her sleuth-quest to speak for the soul beauty of those without a voice. Maisie also wants to avoid what an astute woman, Mrs. Singh, an English woman married to an Indian, tells her about Usha, that painful drawing away when deep feelings are not matched.[63] Aphrodite offers no permanent relationship. Maisie's lover, James Compton, wants just that—marriage—while Maisie feels impelled to travel, like Usha, to leave everything most loved.

Aphrodite's darkness maims and ultimately extinguishes Usha when she is effectively sold by venal Rev. Griffith into the household of Jesmond Martin as both servant and healer for his wife. While Maisie remarks that Griffith helped Jesmond, the prince, find his Cinderella, the resemblance of the fairy tale to the myth of Eros and Psyche reminds us of the harsh price of love.[64] Minus the intervention of a controlling Zeus, Usha comes to be the threat to Robert Martin/Payton of a second father deserting him.

Central to *Leaving Everything Most Loved* is a complex antagonism and attempted reconciliation between Aphrodite, heartbreaker, and Hestia, divinity of home and hearth. Maisie is driven to travel, paradoxically, to find home, a home in herself.[65] She cannot yet say yes to James, for she does not feel at home as partner of a wealthy aristocrat. She does not want to offer him Aphrodite's indifference to his deepest needs. Jesmond Martin fails to win Usha in India because he shows no respect for her hearth, the sacred customs of her family and country. He was the stranger at the hearth who violated it and so was sent away.

In turn, to honor her indigenous hearth, Usha leaves her "home" in England by abruptly departing her employers who love her. She enters the hostel which pretends to offer indigent Indian women a home, while actually exploiting their labor. Maisie finally exposes the pretense of Hestia at the Paige's hostel for the near slavery it is. Betraying Hestia is killing to the psyche.

> "You haven't killed someone, Mr Paige, but you know how to take a life."[66]

Hestia is also vital to the story in images of a web or hidden threads connecting characters just as Hestia sits at the hearth, her connecting fibers making a psychic home of place and city and country. Usha is Hestian not only in refusing to betray her ancestral hearth but also in constellating a divine connectedness in London, where for her the conjunction of Hestia and Aphrodite is fatal. For in the tortured being of the boy, Robert, Usha's apparent power to destroy *his* hearth, his fragile home, with a sick mother and loving stepfather, this second loss of the sacred hearth cannot be borne.

Hestia is the centering homing goddess. In linguistic terms, Hestia gives truth a home in language. Here she exists in tension with tricky Hermes, who makes words deceitful. As soon as Maisie intuits the Hestian nature of the tragedy, she finds Robert threatening yet another hearth. He has a group of poor children at gunpoint when Maisie arrives. While she is prepared to risk the sacrifice of her life to save them, it is Hestian forces that rescue them all. Not only does the children's dog, Nelson, risk his life, but Billy, Maisie's assistant to whom she has developed a familial connection, also steps in.

This is almost Billy's last act for Maisie, as he takes a new job, for the sake of his own family hearth and at Maisie's urging. Maisie's daring rescue enables her to center on the tragic truth of Hestia betrayed by an Aphrodite-like, because divinely inevitable, new relationship between Jesmond Martin and Usha. Yet two other goddesses make important contributions to Maisie's sleuthing. Between all four, Aphrodite, Hestia, Athena and Artemis, she finds her sleuthing no longer satisfying enough, as she tells a policeman.[67] And so her quest changes.

Athena has been a core drive in Maisie's detecting. Basil Khan, wise old friend of her mentor, Maurice, shows her how a light touch on fabric enables the feel of an anomalous thread, "a slub in the weave."[68]

Goddess of weaving Athena, Maisie must tread carefully to detect flaws in her work, those that betray the goddess's essential fidelity to the health of the city or community. Maisie does Athena's work by tracing the tears and remaking of the fabric of the social bond. This includes giving the voiceless a say in their struggle for justice, as Athena gives Orestes a hearing in *The Eumenides*.

Yet Maisie's impulse to detect for Athena the silenced and the social body was ruptured in her previous case in *Elegy for Eddie* (2012).[69] There she discovered that mighty John Otterburn had become so obsessed with preparing Britain for another war with Germany that he was prepared to manipulate the news media and remove by force anyone who got in the way of his plans. Gentle, innocent Eddie became caught up in forces he could not comprehend. He was ruthlessly dispatched, and Maisie is unable to hold Otterburn to account for the death. The voiceless was not heard, and Maisie feels compromised.

> She could no longer continue to put herself forward as a woman of good conduct in her business, when she had been complicit— in her estimation—in keeping secret the role played by John Otterburn in what amounted to murder, even though her voice had been a small one.[70]

Here is the dilemma of the sleuth and the goddess: a sole human may be unable to enact what the goddesses offer: the restoration of the social contract. Put another way, as all hardboiled detectives know (Maisie is not one of them), a single sleuth cannot eradicate endemic corruption, cannot defeat political and monetary power. The main reason why Maisie cannot pursue Otterburn is the looming of World War Two. Polytheism as a lens for social criticism reveals dangerous misalignments between Ares and Aphrodite as well as the curtailing of Athena's strategies for communal justice. Now Hestia is the object of a quest, not an achieved psychological home.

So the way is Artemis. Maisie finds the sleuth role "not enough" because it does not enable her to live her goddesses in the world as it is. She does not want to include the cruelty of Aphrodite in making the wrong marriage to devoted James. She cannot fully restore her community because this war trauma is not sufficiently susceptible to Aphrodite, and her Athena cannot find a place for the Furies in the war to come. Dark forces of the Furies wreck the Hestian hearth that is evoked in the families of *Leaving Everything Most Loved*. Taming them is finally not reconcilable in the sleuth role for Maisie. She has to revert to an even older mythical pattern, that of a quest for the unknown self or as another mentor puts it, "to find escape from the flame of separateness."[71]

Maisie discovers Artemis in Usha Pramal, the victim whose body heals by touch, by intuitive summoning of the powers of nature that knows no boundaries between the human and non-human. Leaving England is certainly leaving everything most loved for Maisie. She is told that it will break her heart open.[72] Yet, such heartbreak finds within a jewel that opens the heart to the making of self. She departs as Artemis in the sense of freeing herself from everything she has known that defined her in a world whose "war to end all wars" has ravaged nature and human nature. And still this war has not ended all wars.

By such an ending in which the sleuth departs from her role, I suggest not only the surmounting of generic forms, such as challenging the role of the fictional detective, but also a resistance to the culture of war, and legacy of war on nature, that has so dominated this twenty-first century so far. *Leaving Everything Most Loved* ultimately reminds us of the price of not leaving the old forms that haunt us in prejudices and hatred of the Other. Far more than most mysteries, this novel offers love without appropriating it.

As Artemis, Maisie leaves alone to find (her) nature. As Artemis, she re-orients her relations to her employees in a less hierarchical and more "sister goddess" connection. The next chapter in Maisie's story will be different. That is what *Leaving Everything Most Loved* offers its readers.

Epilogue for *The Sleuth and the Goddess*

These final eight case histories provide a female hero for the return of the goddesses to a twenty-first century beset by the consequences of psychological, political, and cultural division. Maybe such fear of the Other rests finally on the terror of death. Polytheism is one response to such fear, in that it provides stories of nature and human nature that incorporate the Other in many more forms than dualism. Death is part of the polytheistic universe;

even Zeus cannot reverse its necessity. Here divine beings not subject to death offer stories that incorporate its presence as a necessary part of the deepest energies of life.

If one binary structure of life versus death can be less absolute, then our cultural ways of defining ourselves against an "other" such as other gender, other nature, other culture, start to mutate, adapt, re-story, to restore. So these eight novels tell a story of the goddesses enabling the woman sleuth to get beyond simple opposition to the Other. Conquest or defeat of the Other as another culture, or non-human nature, are not viable options to post 9/11 war or climate change. Rather, the goddesses offer heroism in plural, nuanced forms of home-making, including the ecological work of earth as home, being undivided from non-human nature, erotic bonding, and reason in service of the community that includes hearing the voices of the most vulnerable.

Carrot Cake Murder tells of a fatal loss of innocence as Lake Eden suffers murder of one of its straying members. Fortunately, Hannah Swensen rises Aphrodite-like from the waters in a divine renewal of her and Lake Eden's virginity by naming, and therefore expelling, the culprit. By contrast, *Liberty Falling* is astonishingly prophetic of the 9/11 attacks in its 1999 tale of terrorists at the iconic statue. Here Artemisian Anna Pigeon is a setting for perceiving nature's ultimate triumph over the human-made world. Can she imagine a natural being that spans nature and human nature?

Family figuration leads in *Three Day Town* as southerners Deb and Dwight stumble on a New York murder. Here the bringer of death is invisible and very much within the structure of the city, just as Aphrodite provokes an invisible young girl to cause trouble for Deb's nephew. Meanwhile, *Roast Mortem* looks at playing with fire, a natural force that is essential to life and passion yet is also destructive and killing. Here, life and death give birth to each other as Clare Cosi comes to recognize the trauma of 9/11 still haunting the city.

Mimicry of the ethnic other comes to be the vital clue in *Death Dance*. Alex Cooper and her detective knights end up in an Islamic dome, facing death by entirely indigenous American greed. Again, mis-understanding of the Other is revealed by death dealers from within families and institutions iconic of New York identity. In a change of tone, comedy accommodates death and the Other in *Fearless Fourteen*, in an ethic of survival that adapts Nature and human nature to work with necessity in creating "homegrown security" and its indefatigable potato rocket.

In *Breakdown*, the sleuth is tired and in danger of being inflamed into a fight to the death with forces of capitalist hegemony and ethnic intolerance. It is only by making connections with the "other" as lunatic and as tricked young girl that V.I. can overcome divisions that have threatened to sever her from her own nature. And finally, in *Everything Most Loved*, Maisie Dobbs initiates a richly complex relationship with a victim from another continent. In her leaving, at last, for India, she shows the way to begin to stop fearing other cultures, nature, and death. For to pursue this quest for the goddesses is the essential way to come home, to ourselves and to the planet.

NOTES

The major works of C. G. Jung are referred to in the following notes as:

CW, The Collected Works of C. G. Jung, Volumes 1 to 18. Trans. R. F. C. Hull, ed. by Sir Herbert Read, Michael Fordham, Gerhard Adler and W. McGuire (Princeton, NJ:Princeton University Press, 1934-1954).

Chapter One
Introduction: Mythical Knowing and Detective Fiction

1. Perhaps the first critic to write of the sacred in detective fiction was the poet, W. H. Auden, "The Guilty Vicarage," *The Dyer's Hand and Other Essays* (London: Faber and Faber Limited, 1963), pp. 146-58.

2. In an earlier book I argued for self-conscious fictionality to be intrinsic to the genre. Susan Rowland, *From Agatha Christie to Ruth Rendell: British Women Writers in Detective and Crime Fiction* (London: Palgrave, 2000).

3. Lindsey Davis's series of twenty novels set in Ancient Rome featured informer, Marcus Didius Falco, beginning in 1989 with *The Silver Pigs* and ending in 2010 with *Nemesis*, according to Davis's official website: http://www.lindseydavis.co.uk/publications/.
She then began a new series with Falco's foster daughter, Albia.

4. Arthur Conan Doyle published stories of the iconic Sherlock Holmes and Dr. Watson between 1887 and 1927.

5. Jung, "Commentary on 'The Secret of the Golden Flower,'" 1929, in *Alchemical Studies* CW13, § 54.

6. Christine A. Jackson, *Myth and Ritual in Women's Detective Fiction* (New York: McFarland & Co, Inc., 2002).

7. *Ibid.*, p. 12.

8. Joseph Campbell, *The Hero With a Thousand Faces* (New York: Pantheon Books, 1949).

9. *Ibid.*, p. 167.

10. See Susan Rowland, *The Ecocritical Psyche: Literature, Evolutionary Complexity and Jung* (London and New York: Routledge, 2012), pp. 101-26.

11. Lewis Hyde, *Trickster Makes This World: Mischief, Myth and Art* (New York: Farrar, Straus and Giroux, 1998).

12. See *Aesop: The Complete Fables*, trans. Robert Temple (New York and London: Penguin Classics, 1998).

13. Dorothy L. Sayers, *Strong Poison* (New York: Harper Paperbacks reissue, 2012), p. 132.

14. See Stephen's Knight's persuasive historical account of crime and detective fiction, *Crime Fiction 1800 – 2000: Detection, Death, Diversity* (New York and London: Palgrave, 2004).

15. The best historical account of the emergence of psychoanalysis remains Henri F. Ellenberger, *The Discovery of the Unconscious: The History and Evolution of Dynamic Psychiatry* (New York: Basic Books, 1981).

16. A useful overview and introduction is provided by Christopher Butler, *Modernism: A Very Short Introduction* (Oxford: Oxford University Press, 2010).

17. The first Sherlock Holmes story by Arthur Conan Doyle, "A Study in Scarlet," was published in *Beeton's Christmas Annual* in 1887. See *The Complete Sherlock Holmes: All Four Novels and 56 Short Stories* (New York: Bantam Classics, 1986), pp. 15-87.

18. *Ibid.*, pp. 17-18.

19. Arthur Conan Doyle's short novel, *The Hound of the Baskervilles*, was originally serialized in *The Strand Magazine* 1901-2 and then published as a novel in 1902 by George Newness, London. In *The Complete Sherlock Holmes,* pp. 669-767.

20. James Hillman, "Dionysus in Jung's Writings," first published in *Spring: A Journal of Archetype and Culture*, 1972; *Mythic Figures: Uniform Edition of the Writings of James Hillman, volume 6.1* (Putnam, Connecticut: Spring Publications, Inc., 2007), pp. 15-30.

21. *Ibid.*, pp. 26-9.

22. Jung, "On the Nature of the Psyche," in *The Structure and Dynamics of the Psyche*, CW 8, § 352-3.

23. James Hillman, *Revisioning Psychology* (New York: Harper & Row, 1976), p. 26.

24. James Hillman, 'Dionysus in Jung's Writings," p. 26.

25. *Ibid.*

26. Arthur Conan Doyle, "A Study in Scarlet," p. 18.

27. First published in 1764 by William Bathoe, London. See Horace Walpole, *The Castle of Otranto: A Gothic Story* (Oxford: Oxford World Classics, 1996).

28. For a helpful study see Fred Botting, *Gothic* (London and New York: Routledge, 1996).

29. Arthur Conan Doyle, *The Hound of the Baskervilles*, p. 693.

30. *Ibid.*, p. 679.

31. *Ibid.*, pp. 707-8.

32. *Ibid.*, p. 726.

33. *Ibid.*, p. 760.

34. Sigmund Freud famously built his seminal "Oedipus Complex" on his understanding of the play by Sophocles, *Oedipus Rex*, first performed around 429 BCE. See Sophocles, *The Three Theban Plays: Antigone; Oedipus the King; Oedipus at Colonus* (New York and London: Penguin Classics, 1984).

35. For the Oedipus Complex see Sigmund Freud's *The Interpretation of Dreams* first published in 1899, now trans., James Strachey 1955 (New York: Basic Books, 2010), pp. 244-8.

36. *Ibid.*, p. 246.

37. *Ibid.*

38. I first suggested this possibility for detective fiction in my earlier book, *From Agatha Christie to Ruth Rendell: British Women Writers in Detective and Crime Fiction* (London: Palgrave Macmillan, 2001).

39. *Ibid.*, pp. 12-13, 18, 26.

40. For example see Raymond Chandler, *The Big Sleep* (New York: Alfred A. Knoft, 1939) and Dashiell Hammett, *The Maltese Falcon* (New York: Alfred A. Knoft, 1930).

41. Arguably the English "golden age" or "clue-puzzle" began with Agatha Christie's *The Mysterious Affair at Styles* (London: John Lane, 1920), followed by Dorothy L. Sayers's *Whose Body?* (London: Fisher Unwin, 1923).

42. W. H. Auden, "The Guilty Vicarage," pp. 57-8.

43. See also my article, "The Wasteland and the Grail Knight: Gender, Myth and Cultural Criticism in Detective Fiction," *CLUES: A Journal of Detection*, 2010, vol. 27, No. 2.

44. Jung, "Freud and Jung: Contrasts," in *Modern Man in Search of a Soul* (London: Routledge & Kegan Paul, 1933), pp. 132-42, p. 138.

45. Dorothy L. Sayers, *Clouds of Witness* (London: Fisher Unwin, 1926).

46. *Ibid.*, p. 172.

47. *Ibid.*, p. 173.

48. Jung, "Freud and Jung: Contrasts," p. 138.

49. Jung, "On the Nature of the Psyche," in *The Structure and Dynamics of the Psyche, CW* 8, § 358.

50. Jung, "Psychology and Literature," in *Modern Man in Search of a Soul*, pp. 175-99, p. 188-9.

51. Jung, "On the Nature of the Psyche," *CW* 8, § 397-420.

52. Laurence Coupe, *Myth* (New York and London: Routledge, 1997; 2009).

53. *Ibid.*, pp. 9-11.

54. M.M. Bakhtin, *The Dialogic Imagination: Four Essays*, ed. M. Holquist, trans. by C. Emerson and M. Holquist (Austin: University of Texas Press, 1981).

55. I explored the dialogical nature of archetypal images in an earlier book, *Jung as a Writer* (New York and London, 2005).

56. Jung, "On the Relation of Analytical Psychology to Poetry," in *The Spirit in Man, Art and Literature, CW* 15, § 121.

57. Sara Paretsky, *Deadlock* (New York: Ballantine Books, 1984).

58. *Ibid.*, p. 154.

59. *Ibid.*, p. 158.

60. *Ibid.*, p. 37.

61. Ann Baring and Jules Cashford, *The Myth of the Goddess: Evolution of an Image* (New York and London: Vintage, 1991).

62. Nevada Barr, *Ill Wind* (New York: Berkley Books, 1995).

63. *Ibid.*, pp. 46-7.

64. *Ibid.*, p. 175.

Chapter Two
The Goddesses for Women Writers: Gendering the Genre

1. See Bakhtin, *The Dialogical Imagination*; Hillman, *Mythic Figures*.

2. See Baring & Cashford, *The Myth of the Goddess*.

3. See Sandra Kemp & Judith Squires, ed., *Feminisms*, Oxford Readers (New York and Oxford: Oxford University Press, 1997).

4. See an ecological extension of this argument in my book, *The Ecocritical Psyche*.

5. James Hillman, *Re-Visioning Psychology* (New York: Harper & Row, 1976), pp. 25-35.

6. Stephen Knight, *Crime Fiction 1800 – 2000.*

7. Jung, "On the Psychology of the Trickster Figure," in *The Archetypes of the Collective Unconscious, CW* 9i, § 456-488, § 472.

8. Lewis Hyde, *Trickster Makes This World.*

9. *Ibid.,* p. 20.

10. Rowland, *The Ecocritical Psyche,* pp. 101-26.

11. Jung, "The Paradoxa," in *Mysterium Conjunctionis, CW* 14, § 64 and Hillman, "Dionysus in Jung's Writings," (2007), pp. 26-8.

12. Here the trickster element in this genre is worth comparing to John Beebe's important exploration, John Beebe, "The Trickster in the Arts," *San Francisco Jung Institute Library Journal,* vol. 2, no. 2 (1981): 48.

13. Raymond Chandler, "The Simple Art of Murder," *Atlantic Monthly* 174 (December 1944): 53-9. Republished in revised form in *The Art of the Mystery Story: A Collection of Critical Essays,* ed., with a commentary by Howard Haycraft (New York: Simon & Schuster, 1946), pp. 222-37.

14. *Ibid.,* Haycraft, p. 237.

15. See for example Marlowe's treatment of General Sternwood as the sick king in Raymond Chandler, *The Big Sleep* (New York: Alfred A. Knopf, 1939).

16. W. H. Auden, "The Guilty Vicarage," pp. 57-8.

17. Knight, *Crime Fiction 1800 – 2000,* p. xii, pp. 81-9.

18. Marcia Muller, *Trophies and Dead Things* (New York: The Mysterious Press, 1990).

19. Ginette Paris, *Pagan Meditations: The Worlds of Aphrodite, Artemis and Hestia* (Putnam, CT: Spring Publications, 1986), pp. 90-5.

20. *Ibid.,* p. 19.

21. *Ibid.,* pp. 17-20.

22. On the various birth myths of Aphrodite that suggest both her source in Earth Mother and a patriarchal reinterpretation, see Paris, *Pagan Meditations,* pp. 12-14.

23. *Ibid.,* p. 13.

24. *Ibid.,* p. 60.

25. *Ibid.,* p. 41.

26. *Ibid.,* p. 90.

27. *Ibid.*, p. 96.

28. *Ibid.*, p. 8.

29. *Ibid.*, pp. 16-20.

30. Michael Polyani, *The Tacit Dimension* (London: Routledge & Kegan Paul, 1967).

31. Ginette Paris, *Pagan Meditations*, pp. 79-85.

32. Gill Plain, *Twentieth-Century Crime Fiction: Gender, Sexuality and the Body* (Edinburgh University Press, 2001), pp. 29-55.

33. Paris, *Pagan Meditations*, pp. 22-3.

34. *Ibid.*, pp. 61-3.

35. Marcia Muller, *Trophies and Dead Things*, p. 17.

36. *Ibid.*, pp. 170-3.

37. *Ibid.*, p. 36.

38. Paris, *Pagan Meditations*, p. 110.

39. *Ibid.*, p. 141.

40. *Ibid.*, pp. 141-7.

41. *Ibid.*, pp. 120-4.

42. *Ibid.*, p. 115.

43. Laura Lippman, *The Sugar House* (New York: Avon, 2001), p. 306.

44. Sue Grafton, *K is for Killer* (New York: Henry Holt & Co., 1994), p. 354.

45. *Ibid.*, p. 1.

46. Paris, *Pagan Meditations*, p. 122.

47. Sue Grafton, *K is for Killer*, pp. 255-6.

48. *Ibid.*, p. 364.

49. Paris, *Pagan Meditations*, pp. 167-88.

50. *Ibid.*, p. 167.

51. *Ibid.*, p. 168.

52. Leslie Meier's Lucy Stone mysteries begin with *Mistletoe Murder* (New York: Kensington, 1991); Mary Daheim begins her bed and breakfast series with *Just Desserts* (New York: Avon, 1991); Annie Darling opens her mystery bookstore for Carolyn Hart with *Death on Demand* (New York: Crimeline, 1987).

53. Paris, *Pagan Meditations*, p. 171.

54. Diane Mott Davidson introduces Goldy Schultz in *Catering to Nobody* (New York: Ballantine Books, 1992).

55. Katherine Hall Page brings Faith Fairchild to New England detecting in *The Body in the Belfry* (New York: Avon Books, 1991), while Isis Crawford offers the sisters, Libby and Bernie, first of all in *A Catered Murder* (New York: Kensington, 2003).

56. Joanne Fluke begins the adventures of romantic spinster Hannah Swensen with *Chocolate Chip Cookie Murder* (New York: Kensington, 2003).

57. Sharon McCone's marriage to Hy Ripinsky is announced in *Vanishing Point* (New York: Mysterious Press, 2006).

58. Sue Grafton, *M is for Malice* (New York: Ballantine, 1996).

59. Paris, *Pagan Meditations*, p. 181.

60. Isis Crawford, *A Catered Thanksgiving* (New York: Kensington, 2010).

61. *Ibid.*, p. 1.

62. *Ibid.*, p. 305.

63. *Ibid.*, p. 335.

64. Maureen Murdock, *The Heroine's Journey* (Boston and London: Shambhala, 1990), p. 35.

65. Christine Downing, *Women's Mysteries* (New Orleans: Spring Journal Books, 1992/2003), p. 196.

66. *Ibid.*, p. 33.

67. *Ibid.*, p. 175.

68. Linda Fairstein, *Night Watch* (London and USA: Penguin Group, 2012), p. 54.

69. James Hillman, *Mythic Figures*, p. 55.

70. Linda Fairstein, *Entombed* (New York: Scribner, 2004); *Bad Blood* (New York: Scribner, 2007).

71. Fairstein, *Night Watch*, p. 98.

72. Hillman, *Mythic Figures*, p. 67.

73. *Ibid.*, p. 66.

74. Jacqueline Winspear, *A Lesson in Secrets* (New York: HarperCollins, 2011).

75. Sue Grafton, *O is for Outlaw* (New York: Ballantine, 199).

76. The iconic maverick male policeman is well represented by Inspector Morse who begins for Colin Dexter with *Last Bus to Woodstock* (London: Macmillan, 1975).

77. Hillman, *Mythic Figures*, p. 79.

78. Linda Fairstein, *The Deadhouse* (New York: Scribner, 2001).

79. *Ibid.*, p. 1.

80. Hillman, *Mythic Figures*, p. 64.

81. Linda Fairstein, *The Deadhouse*, p. 178.

Chapter Three
Hestia: Detecting Hearth and Home

1. Paris, *Pagan Mediations*, pp. 181-3.

2. Hillman, *Mythic Figures*, pp. 268-9.

3. Paris, *Pagan Meditations*, pp. 174-88.

4. Hillman, *Mythic Figures*, p. 272.

5. *Ibid.*, p. 235.

6. *Ibid.*

7. Joanne Fluke, Laura Levine, Leslie Meier, *Gingerbread Cookie Murder* (New York: Kensington, 2010), p. 38.

8. Joanna Carl's Lee McKinney mysteries begin with *The Chocolate Cat Caper* (New York: Signet, 2002).

9. Joanna Carl, *The Chocolate Bridal Bash* (New York: Signet, 2006).

10. Georgette Heyer, *Footsteps in the Dark* (Great Britain: Longmans, Green & Co. Ltd, 1932; repr. London: Arrow Books, 2006).

11. Lindsey Davis, *A Body in the Bathhouse* (United Kingdom: Century Random House, 2001).

12. Lindsey Davis, *The Ides of April* (Great Britain: Hodder & Stoughton, 2013).

13. Agatha Christie, *4.50 From Paddington* (UK, London: Collins, 1957).

14. *Ibid.*, p. 54.

15. Nancy Pickard, prolific and renowned author of mysteries began a cooking series with characters invented by Virginia Rich with *The 27 Ingredient Chili Con Carne Murders* (New York: Delacorte Press, 1992).

16. Katherine Hall Page, *The Body in the Gallery* (New York: HarperCollins, 2008), p. 5.

17. Joanne Fluke, *Cream Puff Murder* (New York: Kensington, 2009), p. 134.

18. Diane Mott Davidson, *Dark Tort* (New York: HarperCollins, 2006) 61.

19. *Ibid.*, p. 72.

20. Diane Mott Davidson, *The Main Corpse* (New York: Bantam, 1996).

21. *Ibid.*, p. 267.

22. *Ibid.*, p. 294.

23. *Ibid.*, p. 296.

24. *Ibid.*, p. 232.

25. Sue Grafton, *N is for Noose* (New York: Henry Holt & Co. Inc., 1998).

26. Sara Paretsky, *Toxic Shock* (New York: Delacorte Press, 1988).

27. Pseudonym of academic Carolyn Heilbrun, Amanda Cross introduced the redoubtable scholar sleuth, Kate Fansler with *In the Last Analysis* (New York: Macmillan, 1964).

28. Amanda Cross, *A Trap for Fools* (United States: E.P. Dutton, 1989), p. 152.

29. *Ibid.* p. 153-4.

30. Josephine Tey, *The Daughter of Time* (London: Peter Davies, 1951).

31. *Ibid.*, p. 186.

32. *Ibid.*, p. 188.

33. In Joanne Fluke's *Plum Pudding Murder* (New York: Kensington, 2009).

34. Joanne Fluke, *Cinnamon Roll Murder* (New York: Kensington, 2012).

35. Lindsey Davis, *A Dying Light in Corduba* (London: Century Books, 1996).

36. Lindsey Davis, *Time to Depart* (London: Century Books, 1995).

37. Carolyn Hart, *A Scandal in Fairhaven* (New York: Bantam, 1994).

38. Imogen Quy begins for Jill Paton Walsh with *The Wyndham Case* (London: Hodder & Stoughton, 1993).

39. Jill Paton Walsh, *The Bad Quarto* (London: Hodder & Stoughton, 2007).

40. Jill Paton Walsh, *Debts of Dishonour* (London: Hodder & Stoughton, 2006).

41. Mary Daheim, *Snow Place to Die* (New York: Avon, 1998).

42. Janet Evanovich, *Seven Up* (New York: St. Martin's Press, 2001).

43. Janet Evanovich, *Fearless Fourteen* (New York: St. Martin's Press, 2008), p. 1.

44. Janet Evanovich, *Seven Up* (New York: St. Martin's Press, 2001), p. 7.

45. Knight, *Crime Fiction 1800–2000*, p. xii, pp. 81-9.

46. The development of the genre by the four early British authors is explored in my earlier book, *From Agatha Christie to Ruth Rendell: British Women Writers in Detective and Crime Fiction* (London: Palgrave, 2000).

47. Agatha Christie, *The Mysterious Affair at Styles* (London: Lane, 1920).

48. Dorothy L. Sayers, *Whose Body?* (London: Unwin, 1923).

49. Margery Allingham, *The Crime at Black Dudley* (London: Jarrolds, 1929).

50. Ngaio Marsh, *A Man Lay Dead* (London: Bles, 1934).

51. Jacqueline Winspear's wonderful series open with *Maisie Dobbs* (New York: Soho Press, 2003).

52. Ngaio Marsh, *Final Curtain* (London: Collins, 1947).

53. Marcia Muller, *Listen to the Silence* (New York: Mysterious Press, 2000).

54. Laura Lippman, *The Sugar House* (New York: Avon Books, 2000).

55. Hillman, *Mythic Figures*, p. 235.

56. Marcia Muller, *Listen to the Silence*, p. 47.

57. *Ibid.*, p. 284.

58. Mary Daheim's Alpine series begins with *The Alpine Advocate* (New York: Ballantine, 1992); Agatha Christie introduces St. Mary Mead with *The Murder at the Vicarage* (London: Collins, 1930); Carolyn Hart offers Broward's Rock first with *Death on Demand* (New York: Crimeline, 1987).

59. Lindsey Davis's Falco and Helena Justina rescue Flavia Albia in *The Jupiter Myth* (United Kingdom: Century Random House, 2002).

60. Diane Mott Davidson, *The Cereal Murders* (New York: Bantam, 1994).

61. Diane Mott Davidson *Fatally Flaky* (New York: Avon, 2009).

62. *Ibid.*, p. 11.

63. Diane Mott Davidson, *The Main Corpse*, p. 350.

64. Ann Granger, *That Way Murder Lies* (Great Britain: Headline Publishing, 2004).

65. *Ibid.*, p. 374.

66. Toni L.P. Kelner begins her Laura Fleming series with *Down Home Murder* (New York: Kensington, 1993).

67. *Ibid.*, p. 300.

68. So inevitably, Margaret Maron's Deborah Knott series begins with *The Bootlegger's Daughter* (New York: Mysterious Press, 1992).

69. Margaret Maron, *Southern Discomfort* (New York: Mysterious Press, 1993).

70. *Ibid.*, p. 15.

71. Dorothy L. Sayers, *Strong Poison* (London: Gollancz, 1930).

72. *Ibid.*, p. 45.

Chapter Four
Hunting with Artemis

1. See Paris, *Pagan Meditations*, pp. 109-63; Hillman, *Mythic Figures*, pp. 200-76; Downing, *Women's Mysteries*, pp. 81-143, p. 81.

2. Downing, *Women's Mysteries*, p. 24.

3. Paris, *Pagan Meditations*, pp. 116-8.

4. Downing, *Women's Mysteries*, p. 140.

5. Hillman, *Mythic Figures*, pp. 151-2.

6. *Ibid.*, p. 146.

7. *Ibid.*

8. Dorothy L. Sayers, *Have His Carcass* (London: Gollancz, 1932), p. 9.

9. Joanne Fluke, *Cinnamon Roll Murder* (New York: Kensington, 2012).

10. Sara Paretsky, *Hard Time* (New York: Random House, 1999).

11. Arthur Conan Doyle, *The Hound of the Baskervilles*, 1902 (repr. London: Headline, 2006).

12. V.I. Warshawski's romance with cop, Conrad Rawlings ends in *Tunnel Vision* (New York: Delacorte, 1994).

13. Laurie R. King, *Night Work* (New York: Bantam, 2000).

14. Marcia Muller, *The Ever-Running Man* (New York: Grand Central Publishing, 2007).

15. See Maisie Dobbs in Jacqueline Winspear's *Elegy for Eddie* (New York: HarperCollins, 2012).

16. Marcia Muller, *Vanishing Point* (New York: Mysterious Press, 2006).

17. Nancy Pickard, *The 27 Ingredient Chilli Con Carne Murders* (New York: Delacorte Publishing, 1993), p. 19.

18. Diane Mott Davidson, *The Cereal Murders* (1994); Joanne Fluke, *Cinnamon Roll Murder* (2012).

19. Janet Evanovich, *Two for the Dough* (New York: Scribner, 1996), p. 103.

20. Nevada Barr, *Track of the Cat* (New York: G. P. Putnam's Sons, 1993).

21. *Ibid.*, p. 20.

22. Nevada Barr, *Winter Study* (New York: G. P. Putnam's Sons, 2008).

23. *Ibid.*, p. xi.

24. See note 13.

25. Nevada Barr, *Winter Study*, p. 296.

26. *Ibid.*, p. 456.

27. *Ibid.*, p. 125.

28. *Ibid.*, p. 436.

29. See Jacqueline Winspear, *Pardonable Lies* (United States: Henry Holt & Company, 2005), pp. 4-7.

30. Jacqueline Winspear, *Birds of a Feather* (United States: Soho Press, Inc., 2004).

31. *Ibid.*, pp. 275-83.

32. Sara Paretsky, *Toxic Shock,* first published as *Blood Shot* (New York: Delacorte Press, 1988).

33. Agatha Christie, *Sleeping Murder* (London: Collins, 1976).

34. *Ibid.*, p. 26.

35. Agatha Christie, *The Body in the Library* (London: Collins, 1942).

36. Agatha Christie, *A Pocket Full of Rye* (London: Collins, 1954).

37. *Ibid.*, p. 89.

38. Laurie R. King's *Night Work* (New York: Bantam, 2000).

39. *Ibid.*, p. 375.

40. Linda Fairstein, *The Deadhouse* (New York: Scribner, 2001); *Entombed* (New York: Scribner, 2004); *Killer Heat* (New York: Doubleday, 2008).

41. Fairstein, *Killer Heat*, p. 429.

42. Fairstein, *Entombed*, p. 431.

43. Lauren Henderson, *Chained* (United Kingdom: Hutchinson, 2000).

44. *Ibid.*, p. 6.

45. *Ibid.*, p. 160.

46. *Ibid.*

47. Lindsey Davis, *Three Hands in the Fountain* (United Kingdom: Century, 1997).

48. *Ibid.*, p. 120.

49. Janet Evanovich, *Hot Six* (New York: St Martin's Press, 2000).

50. *Ibid.*, p. 2.

51. Lauren Henderson, *Freeze My Margarita* (United Kingdom: Hutchinson, 1998).

52. *Ibid.*, p. 139.

53. *Ibid.*, p. 247.

54. Paris, *Pagan Meditations*, pp. 109-57.

55. Margery Allingham, *Death of a Ghost* (London: Heinemann, 1934).

56. Sue Grafton, *K is for Killer* (New York: Henry Holt & Co., 1994).

57. Dorothy L. Sayers, *Clouds of Witness* (London: Fisher Unwin, 1926); *Strong Poison* (London: Gollancz, 1930).

58. Dorothy L. Sayers, *Busman's Honeymoon* (London: Gollancz, 1937), pp. 394-7.

59. Dorothy L. Sayers, *Gaudy Night* (London: Gollancz, 1935).

60. *Ibid.*, p. 34.

61. A stray dog, later named Nux, saves Falco in Lindsey Davis's *Time to Depart* (United Kingdom, Century, 1995).

62. Nux gives birth in Lindsey Davis's *Ode to a Banker* (United Kingdom: Century, 2000), p. 200.

63. Nevada Barr, *Ill Wind* (New York: Berkley Books, 1995), p. 22.

64. Marcia Muller, *Where Echoes Live* (New York: Mysterious Press, 1991).

65. *Ibid.*, p. 304.

66. *Ibid.*, p. 96.

67. *Ibid.*, p. 97-8.

68. *Ibid.*, p. 98.

69. See note 63.

70. Sue Grafton, *C is for Corpse* (New York: Ballantine, 1986); *M is for Malice* (New York: Ballantine, 1996).

71. See Faith Fairchild for Katherine Hall Page, *The Body in the Gallery* (New York: HarperCollins, 2008), p. 5.

72. Lucy Stone puts detecting above domestic ties in numerous works but particularly in *Back to School Murder* (New York: Kensington, 1997), where she meets a flirtatious college teacher, p. 240-2.

73. The Furies are tamed by the goddess Athena in Aeschylus's play, "The Eumenides" in *The Oresteia*, trans. Robert Fagles (New York: Penguin Books, 1966), pp. 227-77.

Chapter Five
Athena's Justice

1. Sophocles, *The Three Theban Plays: Antigone; Oedipus the King; Oedipus at Colonus* (New York and London: Penguin Classics, 1984).
2. Aeschylus, "*The Eumenides,*" in *The Oresteia*, trans. Robert Fagles (New York: Penguin Books, 1966), p. 239, l. 183-4.
3. *Ibid.*, p. 249, l. 428.
4. *Ibid.*, p. 269, l. 860-1.
5. *Ibid.*, p. 268, l. 839-41.
6. Hillman, *Mythic Figures*, pp. 55-72.
7. *Ibid.*, p. 66.
8. *Ibid.*
9. *Ibid.*, p. 69.
10. *Ibid.*, p. 70.
11. *Ibid.*, p. 68.
12. Paris, *Pagan Meditations*, p. 162.
13. Downing, *Women's Mysteries*, p. 33.
14. Aeschylus, *The Eumenides*, p. 270, l. 904.
15. Margaret Maron's Deborah Knott series begins with *The Bootlegger's Daughter* (New York: Mysterious Press, 1992).
16. Aeschylus, *The Eumenides,* p. 262, l. 709-10.
17. Hillman, *Mythic Figures*, p. 67.
18. *Ibid.*, p. 65.
19. Ngaio Marsh, *Final Curtain* (London: Collins, 1947).
20. Ngaio Marsh, *Opening Night* (London: Collins, 1951).
21. Ngaio Marsh, *Spinsters in Jeopardy* (London: Collins, 1954).
22. Ngaio Marsh, *Off With His Head* (London: Collins, 1957).
23. Aeschylus, *The Eumenides*, p. 251, l. 450-1.
24. Patricia Cornwell begins her Kay Scarpetta novels with *Postmortem* (New York: Simon & Schuster, 1990).
25. Jill Paton Walsh, *A Piece of Justice* (Great Britain: Hodder & Stoughton, 1995), p. 31.
26. Jill Paton Walsh, *Debts of Dishonor* (Great Britain: Hodder & Stoughton, 2006).

27. *Ibid.*

28. Michelle Spring, *Nights in White Satin* (London: Orion Books, 1999).

29. Mary Daheim begins her bed and breakfast series with *Just Desserts* (New York: Avon, 1991) and the Alpine series begins with *The Alpine Advocate* (New York: Ballantine, 1992).

30. Kate Charles began her clerical mysteries featuring David Middleton Brown and Lucy Kingsley with *A Drink of Deadly Wine* (London: Headline, 1991).

31. Aeschylus, *The Eumenides,* p. 262, l. 709-10.

32. Kate Charles, *Appointed to Die* (London: Headline, 1993).

33. *Ibid.,* p. 333.

34. *Ibid.,* p. 179.

35. *Ibid.,* p. 1.

36. Diane Mott Davidson, *The Whole Enchilada* (New York: HarperCollins, 2013), pp. 274-5.

37. Paris, *Pagan Meditations,* p. 162.

38. Marcia Muller, *Wolf in the Shadows* (New York: Warner Books, 1993).

39. *Ibid.,* p. 76.

40. *Ibid.,* p. 124.

41. Laurie R. King, *Night Work* (New York: Bantam, 2000).

42. Diane Mott Davidson, *The Main Corpse* (New York: Bantam, 1996), p. 132.

43. Jill Paton Walsh, *The Attenbury Emeralds* (New York: St Martin's Press, 2010).

44. Dorothy L. Sayers, *Clouds of Witness* (London: Fisher Unwin, 1926).

45. Dorothy L. Sayers, *Gaudy Night* (London: Gollancz, 1935).

46. *Ibid.,* p. 74.

47. *Ibid.,* p. 284.

48. *Ibid.,* p. 427.

49. *Ibid.,* p. 420.

50. Ruth Rendell begins her Wexford novels with *From Doon with Death* (London: Hutchinson, 1964).

51. Ruth Rendell, *Road Rage* (London: Hutchinson, 1997).

52. Ruth Rendell, *Simisola* (London: Hutchinson, 1995).

53. Carolyn Hart, *Dead Man's Island* (New York: Bantam, 1993).

54. Margaret Maron, *Christmas Mourning* (New York: Grand Central Books, 2020), p. 52.

55. Amanda Cross, *A Death in the Faculty* (New York: E.P. Dutton, 1981).

56. *Ibid.*, p. 145.

57. *Ibid.*, p. 146.

58. *Ibid.*, p. 148.

59. Sara Paretsky, *Breakdown* (New York: Penguin Books, 2012), p. 177.

60. Agatha Christie, *Murder on the Orient Express* (London: Collins, 1934).

61. Hillman, *Mythic Figures*, p. 72.

62. Ruth Rendell, *The Veiled One* (UK: Hutchinson, 1988).

63. Barbara Hambly, *Dead Water* (New York: Bantam, 2004).

64. *Ibid.*, p. 147.

65. *Ibid.*, p. 368.

66. *Ibid.*, p. 292.

67. Laurie R. King, *To Play the Fool* (New York: HarperCollins, 1995).

68. Carolyn Hart, *Mint Julep Murder* (New York: Bantam, 1995).

Chapter Six
The Mysteries of Aphrodite

1. See Chapter 1 and also Chapter 5 of my book, *The Ecocritical Psyche* (New York: Routledge, 2012), "Hunting Signs with the Trickster Detectives," pp. 101-26.

2. See Chapter 7 of *Jung as a Writer* (New York: Routledge, 2005), "Culture, Ethics, Synchronicity and the Goddess," pp. 171-95.

3. Ann Baring and Jules Cashford, *The Myth of the Goddess: Evolution of an Image* (New York and London: Vintage, 1991).

4. Paris, *Pagan Meditations*, p. 19.

5. Downing, *Women's Mysteries*, p. 197.

6. *Ibid.*, p. 92.

7. Paris, *Pagan Meditations*, pp. 16-30.

8. Linda Fairstein, *Cold Hit* (New York: Scribner,1999), p. 31.

9. Cleo Coyle, *Murder by Mocha* (New York: Berkley Prime Crime, 2011).

10. Hillman, *Mythic Figures*, p. 220.

11. Cleo Coyle, *Murder by Mocha*, p. 284.

12. *Ibid.*, p. 202.

13. Sara Paretsky, *Toxic Shock* (New York: Delacorte Press, 1988).

14. Paris, *Pagan Meditations*, p. 14.

15. *Ibid.*

16. Toni L. P. Kelner, *Blast from the Past* (New York: Berkley Prime Crime, 2011), p. 278.

17. Hillman, *Mythic Figures*, pp. 203, 220.

18. Paris, *Pagan Meditations*, p. 24.

19. Cleo Coyle, *Latte Trouble* (New York: Berkley Prime Crime, 2005), p. 161.

20. Janet Evanovich, *Four to Score* (New York: St Martin's Press, 1998).

21. *Ibid.*, p. 85.

22. Laura Lippman, *The Girl in the Green Raincoat* (New York: William Morrow, 2008).

23. *Ibid.*, p. 47.

24. *Ibid.*, p. 48.

25. *Ibid.*, p. 115.

26. *Ibid.*, p. 154.

27. *Ibid.*

28. The best account of the role of Aphrodite in the Trojan War is to be found in the retelling of Homer's Iliad by Lindsay Clarke, *The War at Troy* (Great Britain: HarperCollins Publishers, 2004).

29. Lindsey Davis, *The Ides of April* (Great Britain: Hodder & Stoughton, 2013).

30. Sara Paretsky, *Bitter Medicine* (New York: William Morrow, 1987).

31. Joanne Fluke, *Cream Puff Murder* (New York: Kensington Books, 2009).

32. See note 10.

33. Apuleius, E. J. Kenney, trans., *The Golden Ass* (New York: Penguin Classics, 1998).

34. Erich Neumann, *Amor and Psyche* (Princeton: Princeton University Press, 1956).

35. Kate Charles, *A Drink of Deadly Wine* (London: Headline, 1991).

36. Sara Paretsky, *Burn Marks* (New York: Delacorte Press, 1990).

37. Linda Fairstein, *Night Watch* (London and USA: Penguin Group, 2012).

38. Jacqueline Winspear, *Elegy for Eddie* (New York: Harper-Collins, 2012).

39. *Ibid.*, p. 29.

40. *Ibid.*, p. 293.

41. *Ibid.*

42. *Ibid.*, p. 140.

43. *Ibid.*, p. 8.

44. *Ibid.*, p. 104.

45. Agatha Christie, *Appointment with Death* (London: Collins, 1938).

46. Lindsey Davis, *See Delphi and Die* (London: Century Books, 2005).

47. Nevada Barr, *Winter Study* (New York: G. P. Putnam's Sons, 2008).

48. Diane Mott Davidson, *The Main Corpse* (New York: Bantam, 1996).

49. Barbara Hambly, *Fever Season* (New York: Bantam, 1998).

50. Laurie R. King, *Night Work* (New York: Bantam, 2000).

51. Marcia Muller, *Where Echoes Live* (New York: Mysterious Press, 1991).

52. Cleo Coyle, *Roast Mortem* (New York: Berkley Prime Crime, 2010), p. 252.

53. Diane Mott Davidson, *The Last Suppers* (New York: Bantam, 1994).

54. Marcia Muller, *City of Whispers* (New York: Grand Central Publishing, 2011).

55. Marcia Muller, *Looking for Yesterday* (New York: Grand Central Publishing, 2012).

56. Marcia Muller, *Coming Back* (New York: Grand Central Publishing, 2010).

57. *Ibid.*, p. 95.

58. Muller, *Looking for Yesterday*, p. 293.

59. Antonia Fraser, *Oxford Blood* (Great Britain: Wiedenfeld & Nicholson, 1985).

60. *Ibid.*, p. 26.

61. *Ibid.*, p. 214.

62. Paris, *Pagan Meditations*, p. 68.

63. Carolyn Hart, *Mint Julep Murder* (New York: Bantam, 1995).

64. Fluke, *Cream Puff Murder*, p. 170.

65. Jacqueline Winspear, *Pardonable Lies* (United States: Henry Holt & Company, 2005).

66. *Ibid.*, p. 297.

Chapter Seven
The Nature of the 21st Century: The sleuth and the goddess after 9/11

1. Joseph Campbell, *The Hero With a Thousand Faces* (New York: Pantheon Books, 1949).

2. Ursula Le Guin, "The Carrier Bag Theory of Fiction," in C. Glotfelty and H. Fromm eds., *The Ecocriticism Reader: Landmarks in Literary Ecology* (Athens and London: The University of Georgia Press, 1996), pp. 149-54.

3. Joanne Fluke, *Carrot Cake Murder* (New York: Kensington, 2008).

4. *Ibid.*, p. 99.

5. *Ibid.*, p. 75.

6. *Ibid.*, p. 52.

7. *Ibid.*, p. 311.

8. *Ibid.*, p. 312.

9. Nevada Barr, *Liberty Falling* (New York: HarperCollins Publishers, 1999).

10. *Ibid.*, p. 343.

11. *Ibid.*, p. 89.

12. *Ibid.*, p. 282.

13. *Ibid.*, p. 298.

14. Margaret Maron, *Three-Day Town* (New York: Grand Central Publishing, 2011).

15. *Ibid.*, p. 287.

16. *Ibid.*, p. 224, p. 295.

17. *Ibid.*, p. 280.

18. *Ibid.*, p. 283.

19. *Ibid.*, p. 132.

20. *Ibid.*, p. 94.

21. *Ibid.*, p. 295.

22. Cleo Coyle, *Roast Mortem* (New York: Berkley Prime Crime, 2010).

23. *Ibid.*, p. 254.

24. *Ibid.*, p. 52.

25. *Ibid.*, p. 89.
26. *Ibid.*, p. 9.
27. *Ibid.*, p. 143.
28. *Ibid.*, p. 53.
29. *Ibid.*, p. 149.
30. *Ibid.*, p. 264.
31. Janet Evanovich, *Fearless Fourteen* (New York: St. Martin's Press, 2008).
32. *Ibid.*, p. 206.
33. *Ibid.*, p. 215.
34. *Ibid.*, p. 87.
35. *Ibid.*, p. 253.
36. *Ibid.*, p. 236.
37. *Ibid.*, p. 303.
38. Linda Fairstein, *Death Dance* (New York: Scribner, 2006).
39. *Ibid.*, p. 131.
40. *Ibid.*, p. 133.
41. *Ibid.*, p. 344.
42. *Ibid.*, p. 346.
43. *Ibid.*, p. 351.
44. *Ibid.*, p. 295.
45. *Ibid.*, p. 332.
46. *Ibid.*, p. 482.
47. *Ibid.*
48. Sara Paretsky, *Breakdown* (New York: Penguin Books, 2012).
49. George Eliot, *The Mill on the Floss* (New York: Dover Publications, 1860/2003), p. 424.
50. Paretsky, *Breakdown*, p. 75.
51. *Ibid.*, p. 26.
52. *Ibid.*, p. 93.
53. *Ibid.*, p. 236.
54. *Ibid.*, p. 102.
55. *Ibid.*, p. 295.
56. *Ibid.*, pp. 292-3.
57. *Ibid.*, p. 425.
58. *Ibid.*, p. 431.
59. Jacqueline Winspear, *Leaving Everything Most Loved* (New York: HarperCollins Publishers, 2013).

60. *Ibid.*, p. 269.

61. *Ibid.*, p. 143.

62. Paris, *Pagan Meditations*, pp. 79-85.

63. *Ibid.*, p. 147.

64. *Ibid.*, p. 297.

65. *Ibid.*, p. 171.

66. *Ibid.*, p. 235.

67. *Ibid.*, p. 313.

68. *Ibid.*, p. 116.

69. Jacqueline Winspear, *Elegy for Eddie* (New York: Harper-Collins, 2012).

70. Winspear, *Leaving Everything Most Loved,* p. 276.

71. *Ibid.*, p. 11.

72. *Ibid.*, p. 326.

SUBJECT INDEX

A

Actaeon, 93–94, 96, 112, 124, 129

Agamemnon, 42, 53, 128

Ambivalence: about sister goddess, 105; to institutions, 97; to nymphs, 105–106

Animals: Artemis animation with, 119–120; Artemis as embodiment of, 94; relationships with, 98–99

Animism, 25, 162

Aphrodite: associated with carnal bodily knowledge, 163, 164–166, 169, 190; birth of, 169–171; case histories related to, 166–169, 173–176, 180–183, 187–190; characteristics of, 161–163; destructive nature of, 38, 40–41, 164, 176–178; as Earth Mother, 163; Ero's inheritance from, 171; evoked in *Clouds of Witnesses*, 18, 31; inspiring passion and lust, 168–169, 171–173, 219; multiple dimensions of, 35–39; ordeals of Psyche, 183–187; penchant for adornments, 172; sexual symbolism of, 35–36, 37, 164–166; as trickster, 39; truthfulness of, 190–192

Apollo, 6, 7, 35, 36, 128, 178

Archetypes: Jung's view of, 7, 19–22; threatening female, 198–199

Ares, 35, 38, 39, 176, 183, 204

Artemis: animation of, 114, 119–120, 124; association with virginity, 93, 95, 107, 125; autonomy of, 95–97; avenging women's deaths, 46,

102–103, 122–123; case histories related to, 93, 97, 100–103, 108–111, 115–117, 120–123, 184, 185; characterized in detective fiction, 42–43; evoked in *K is for Killer*, 43–46, 74, 117; as hunter and killer, 102–103, 122–123; hunting alone, 97; Kali vs., 109, 111; knowing through non-human nature, 31; legacy of death and sacrifice, 42, 117–119, 210; nature of, 93–95, 123–125; post-9/11 themes invoking, 198, 199–200, 215–216; punishment of Actaeon, 93–94, 124, 129; as sister goddess, 103–106; soul of gendered embodiment, 111–114, 124; themes in *Trophies and Dead Things*, 41; Titanism and, 94–95, 111, 122, 123–124

Athena: case histories related to, 118, 137–139, 143–146, 148–151, 154–157; as containment, 143, 146–148; Cooper's embodiment of, 53–54, 55, 56–58, 210, 211; embodying lawyer sleuth values, 53, 55; evoked in Warshawski's actions, 31, 214–215; Furies and, 53, 127–129, 157–159; golden apple and, 176; institutions associated with, 134–137; normalizing perspective of, 151–154; as perspective of communal life, 132–134; protecting community, 53–55, 128–129;

Feasting, 91–92

Femininity: Aphrodite's sexual symbolism, 35–36, 37, 164–166; Artemis as incarnation of, 43, 44–46, 93, 111–114; converting dualism into diversity, 32–34; embodied in Athena's themes, 52–55; found in novel form, 194; · Freudian psychoanalysis of masculinity and, 36; Hestian Earth-centered themes, 46–49; individuation of, 178; threatening female archetypes, 198–199; virginity and, 36

Fertility: food and, 66; mysteries of birthing, 94

Fisher King, 15, 34

Food: about cooking cozies, 87–88; as aphrodisiac, 166–167; catering businesses in fiction, 50–52; as hearth-making element, 61–62, 66–68; Hestia's role with, 47–48, 61, 91–92; psychic centering in preparing, 84

Forensics, 6

Furies: Athena's role with, 127–129, 157–159; confronting, 208–209, 210, 211–212, 215–216, 221–222; destructive nature of, 128–129, 135, 136; detectives take place of, 130–132; in *Gaudy Night*, 145–146; as "Kindly Ones", 128, 155, 157; making place for within divine order, 53; revenge by, 128–129, 135, 136

G

Gaia, 170

Gender: converting dualism of into feminine diversity, 32–34; role of Sky Father in creation of, 30–31;

traits of Earth Mother and Sky Father, 30, 163; of tricksters, 161; of writers developing hardboiled and cozy sleuths, 34–35. *See also* Goddesses; Gods

Goddesses: Artemis as sister, 103–106; becoming diseases, 2; creation myths about, 24–26; function of, 194; Kali, 108, 109, 110, 111; as modes of knowing, 5, 31; working in polytheistic cosmos, 60. *See also* *specific goddesses*

Gods: Apollo, 6, 7, 35, 36, 128, 178; Ares, 35, 38, 39, 176, 183, 204; Hephaestus, 39; Hephaistos, 204, 205–206; Hermes pairing with Hestia, 49, 59, 60, 61, 78, 81–83, 85, 90; as modes of knowing, 5, 31; working in polytheistic cosmos, 60; Zeus, 36, 52, 129, 130, 176, 185, 187. *See also* Dionysus; Sky Father

Golden apple, 176–178

Gothic elements, 8, 9, 10–11

Grail. *See* Holy Grail

H

Hades, 45, 56, 181

Hardboiled detective fiction: characteristics of, 15; cozy vs., 86, 152; masculine heroes in, 33–34; women sleuths in, 104–105

Healing: hearth and home, 85–89; as Hestian theme, 65; rebalancing creation myths for, 31; 21st century community, 201, 202, 203, 211–212

Helen, 176

Hephaestus, 39

Hephaistos, 204, 205–206

Hera, 176

Hermes, 49, 59, 60, 61, 78, 81–83, 85, 90, 190

INDEX OF WORKS CITED

AUTHORS INDEX

CHARACTER INDEX

Printed in the United States
by Baker & Taylor Publisher Services